SARVEPALLI RA
acknowledged
philosophical and religious tradition to the West. He was also a statesman of distinction who served as President of India.

Starting his career at the Madras Presidency College in 1909, he was appointed Professor of Philosophy in the University of Mysore in 1918. Three years later, he was appointed to the prestigious King George V Chair of Mental and Moral Science in the University of Calcutta.

The years from 1936 to 1948 saw Radhakrishnan assume many responsibilities — the Spalding Professor of Eastern Religions and Ethics at Oxford University, Fellow of the British Academy, Vice-Chancellor of the Banaras Hindu University, Chancellor, University of Delhi among others.

In may 1962 he took office as the President of India.

'Among the philosophers of our time, no one has achieved so much in so many fields... Never in the history of philosophy has there been quite such a world-figure.'

George P Conger in *The Hindu*

By the same author
in
Orient Paperbacks

Recovery of Faith
The Present Crisis of Faith
the Creative Life
Towards a New World
Living with a Purpose
True Knowledge
Our Heritage
Indian Religions

॥ ३ॐ ॥ श्रीरामः ॥

Essays on
RELIGION, SCIENCE & CULTURE

S. Radhakrishnan

Orient Paperbacks
DELHI | MUMBAI | HYDERABAD

ISBN : 978-81-222-0012-6

Religion, Science & Culture

Subject: Philosophy / Essays

1st Published 1968
This edition 2013

Published by
Orient Paperbacks
(A division of Vision Books Pvt. Ltd.)
5A/8 Ansari Road, New Delhi-110 002
www.orientpaperbacks.com

Cover design by Sandeep Sinha for Vision Studio

Printed at
Ravindra Printing Press, Delhi-110 006, India

CONTENTS

Religion in a Changing World	7
Indian Religious Thought and Civilization	20
The World Communities of Ideals	47
Internationalism: A Challenge & An Oppurtunity	67
The Metaphysical Quest	82
The Asian View	99
Science and Religion	108

The spirit of science leads to the refinement of religion.
Intellectual authority should be treated with respect,
and not merely inherited authority.

CHAPTER 1

RELIGION IN A CHANGING WORLD

The world has been shrinking at an increasing pace with the advance of communications and technology. We have now the physical basis for a unified world community. We do not any more live in separate worlds. Asia and Africa cannot raise the living standards of their peoples without technical support from Europe and America. These latter cannot subsist without the commodities and raw materials of other parts of the world. Besides, science and technology have put great powers in the hands of men, which rightly used can give strength, freedom and better life to millions of human beings, or abused will bring chaos and destruction.

Professor Adrian, President of the Royal Society of England, in his inaugural address on 'Science and Human Nature' at the 116th annual meeting of the British

Association for the Advancement of Science, said that the control achieved over the forces of nature was so complete 'that we might soon be able to destroy two-thirds of the world by pressing a button.' The destructive power now in the hands of men has reached such terrifying proportions that we cannot afford to take any risks. World solidarity, *lokasaṁgraha*, is no more a pious dream. It is an urgent practical necessity. The unity of the world is being shaped through the logic of events, material, economic and political. If it is to endure, it must find psychological unity, spiritual coherence. The world, unified as a body, is groping for its soul. If mankind is to save itself, it must change the axis of its thought and life. There is throughout the world an increasing spread to materialism, and automation.

The need for religion, for a system of thought, for devotion to a cause which will give our fragile and fugitive existence significance and value does not require much elaborate argument. It is an intrinsic element of human nature. The question is, what kind of religion? Is it a religion of love and brotherhood or of power and hate? Secular ideologies ask us to worship wealth and comfort, class or nation. The question is therefore not, religion or no religion, but what kind of religion?

So long as any religious system is capable of responding creatively to every fresh challenge, whether it comes by the way of outer events or of ideas, it is healthy and progressive. When it fails to do so it is on the decline. The breakdown of a society is generally due to a failure to devise adequate responses to new challenges, to a failure to retain the voluntary allegiance of the common people who, exposed to new winds of thought and criticism, are destitute of faith, though afraid of scepticism. Unless religions reckon with the forces at work and deal with them creatively, they are likely to fade away.

We live in an age of science and we cannot be called upon to accept incredible dogmas or exclusive revelations. It is again an age of humanism. Religions which are insensitive to human ills and social crimes do not appeal to the modern man. Religions which make for division, discord and disintegration and do not foster unity, understanding and coherence, play into the hands of the opponents of religion.

* * *

The general impression that the spirit of science is opposed to that of religion is unfortunate and untrue. One of the main arguments for the religious thesis is the objective consideration of the cosmos. What is called natural theology is based on the study of the empirically observable facts and not on authoritative sources such as revelations or traditions. Those who attempt to construct by reasoned argument a theory of ultimate being from a survey of the facts of nature are adopting the scientific method. The *Brahma Sūtra* which opens with the *sūtra, athāto brahmajijñāsā,* now therefore the desire to know Brahman, is followed by the other, *janmādyasya yataḥ,* Brahman is that from which the origin of this world (along with subsistence and dissolution) proceeds. The *sūtra* refers to the account in the third chapter of the *Taittiriya Upaniṣad.* There has been a steady ascent from the inorganic to the organic, from the organic to the sentient, from the sentient to the rational life. The rational has to grow into the spiritual which is as far above the purely rational, as the rational is above the purely sentient. A spiritual fellowship is the meaning of history. The purpose of the cosmic process is the city of God in and out of time. Earth is the seed ground of the new life of spirit. Earth and heaven are intermingled.[1]

The spirit of science does not suggest that the ultimate beginning is matter. We may split the atom. The mind of man which splits it is superior to the atom. The achievements of science stand as witnesses to the spirit in man. The nature of the cosmic evolution with its order and progress suggests the reality of underlying spirit. I need not refer to the metaphysicians trained in science like, Lloyd Morgan, Alexander, Whitehead, and others.

Albert Einstein in his book, *The World As I See It,* observes that the scientist's 'religious feeling takes the form of a rapturous amazement at the harmony of natural law, which reveals an intelligence of such superiority that, compared with it, all the systematic thinking and acting of human beings is an utterly insignificant reflection. This feeling is the guiding principle of his life and work, in so far as he succeeds in keeping himself from the shackles of selfish desire. It is beyond question closely akin to that which has possessed the religious geniuses of all ages.'

Scientists are men dedicated, set apart. They have renounced the life of action. Their life as the pursuit of truth is service of God, who is Truth: *satya svarūpa satyanārāyaṇa.* Erasmus delivered the great dictum: 'Wherever you encounter truth, you look upon Christianity.'

The spirit of science leads to the refinement of religion. Religion is not magic or witchcraft, quackery or

[1] The vision of a renewed creation finds poignant expression in the Apocalypse of St. John: 'Behold the tabernacle of God with men; and He will dwell with them and they shall be His people; and God Himself with them shall be their God.

'And God shall wipe away all tears from their eyes and death shall be no more. Nor shall mourning nor crying nor sorrow be any more, for the former things are passed away. And He that sat on the throne said: "Behold, I make all things new".'

superstition. It is not to be confused with outdated dogmas, incredible superstitions, which are hindrances and barriers, which spoil the simplicity of spiritual life. Intellectual authority should be treated with respect and not merely inherited authority. Besides, science requires us to adopt an empirical attitude. Experience is not limited to the data of perception or introspection. It embraces paranormal phenomena and spiritual states. All religions are rooted in experience.

* * *

Among the relics of the Indus civilization are found figures which are the prototype of Siva, suggesting that he who explores his inward nature and integrates it is the ideal man. This image has haunted the spiritual landscape of this country from those early times till today. The *Upaniṣads* require us to acquire *brahma-vidyā* or *ātma-vidyā.*[2] The *Kaṭha Upaniṣad* says, that man is turned outward by his senses and so loses contact with himself. He has lost his way. His soul has become immersed in outer things, in power and possessions. It must turn round, *āvṛtta-caksuḥ,* to find its right direction and discover the meaning and reality it has lost. The Jina is one who conquers his self. He is the *mahāvíra,* one who has battled with his inward nature and triumphed over it. The Buddha asks us to seek enlightenment, *bodhi.* These different religions ask us to change our unregenerate nature, to replace *avidyā,* ignorance, by *vidyā* wisdom.

[2] Some aspects of Greek religion emphasize self-knowledge. Heraclitus said: 'I sought myself.' The injunction to know the self was written over the porch at Delphi. Socrates started his quest by becoming aware that he does not know himself and indeed that he does not know anything. When we know that we do not know, we begin to know ourselves.

Of course, they do not mean by *vidyā* textual learning. The man who knows all about the texts is *mantravit,* not *ātmavit. Nāyamātmā pravacanena labhyo na medhayā na bahunā śrutena.*[3] 'This self cannot be attained by instruction nor by intellectual power nor through much hearing.'

Religion is not mere intellectural conformity or ceremonial piety; it is spiritual adventure. It is not theology but practice. To assume that we have discovered final truth is the fatal error. The human mind is sadly crippled in its religious thinking by the belief that truth has been found, embodied, standardized and nothing remains for us to do except to reproduce feebly some precious features of an immutable perfection. Religion is fulfilment of man's life, an experience in which every aspect of his being is raised to its highest extent. What is needed is a change of consciousness, a rebornness, an inner evolution, a change in understanding.

The distinction between time and eternity is qualitative. No quantity of time can produce eternity—*nāsty akṛtaḥ kṛtena.* Our thought must be lifted to another order of reality about time.[4] The change from reason to spirit is a qualitative one.

There is no such thing as an automatic evolution of man, something that happens according to the laws of heredity and natural selection. Man's evolution is bound up with his conscious effort. As he is, man is an unfinished being. He has to grow into a regenerate being and permit the currents of universal life to flow through him. Those who have evolved, who have realized their latent possibilities, who are reborn, serve as examples and guides to others.

---

[3] *Kaṭha Upaniṣad,* 1,2.33.

[4] Cf. Spinoza: 'Eternity cannot be defined by time or have any relation to it.'

This is the teaching of Christiantiy. Jesus asks us to bring about this rebirth, the second birth, to become a new man. The change takes place by inner contemplation, not outer life. When Jesus rebukes the Pharisees, he is condemning the man of pretences who keeps up appearances, who conforms to the letter of the law. 'Except your righteousness shall exceed the righteousness of the Scribes and the Pharisees, ye shall in no wise enter into the Kingdom of Heaven.'[5] We must act not from the idea of reward but for the sake of what is good in itself. To attain heaven which is the higher level of understanding of being, one has to undergo inner growth, growth in wisdom and stature through prayer and fasting, through meditation and self-control. Jesus says of John the Baptist that he is the best of those born of women but the least in the Kindgdom of Heaven was greater than he.[6] John speaks to us of salvation through moral life. He tells us what to *do*, not what to *be*. Jesus insists on inner transformation. John symbolizes the man of external piety, Jesus the man of inner understanding.[7] John asks us to become better, Jesus asks us to become different, new. John the Baptist was puzzled when he heard that Jesus and his disciples ate and drank and did not fast. They plucked the ears of corn on the Sabbath day. Jesus healed on the Sabbath day. John is still a man born of woman; he has not experienced rebirth. 'Except a man be born again, he cannot see the Kingdom of God.'[8] The writer to the Ephesians says: 'Awake thou that sleepest and arise from the dead.'[9] We are like dead people; we should wake up. Christian teaching in its origin, before it became externalized and organized, was

---

[5] *Matthew, V,* 20.
[6] *Matthew, III, 2; Luke,* III, 10-14.
[7] *Luke,* XVII, 20.
[8] *John,* III, 3.
[9] *IV,* 14.

about awakening from sleep through the light shed by the inner wisdom. Jesus was one who had awakened and taught others the way of awakening. In this way, says the writer to the Ephesians, 'you will redeem the time.'[10]

The Kingdom of Heaven is the highest state attainable by man. It is within us.[11] 'He hath set eternity in the heart of man.'[12] Man stands between the visible and the invisible world. Our ordinary level of consciousness is not the highest form or the sole mode of experience possible to man. To get at the inner experience we must abstract from the outer. We must get away from the tumult of sense impressions, the riot of thoughts, the surgings of emotions, the throbs of desires. Boehme says that we come into the reality of our being and perceive everything in a new relation, 'if we can stand still from self-thinking and self-willing and stop the wheel of imagination and the senses.' Karl Barth observes: 'Men suffer, because bearing within them an invisible they find this unobservable inner world met by the tangible, foreign, other outer world, desperately visible, dislocated, its fragments jostling one another, yet mightly powerful and strangely meaning and hostile.'[13]

* * *

---

[10] V, 16. William Law following Boehme writes: 'Do but suppose a man to know himself, that he comes into this world on no other errand but to arise out of the vanity of time. Do but suppose him to govern his inward thought and outward action by this view of himself and then to him every day has lost all its evil; prosperity and adversity have no difference, because he receives them and uses them in the same spirit.' *The Works of William Law* (1749), reprinted in 1893, Vol. VII, p. 1.

[11] *John,* III, 3.

[12] *Ecclesiastes,* III, II.

[13] *Commentary on Romans,* p. 306.

The great scriptures are the records of the sayings of the prophets, *āpta-vacana*. We do not prove the truth of an idea by demonstrating that its author lived or that he was a respectable man. The evidence of truth lies in man's experience of it when it enters into him. The Buddha asks us to accept his words after examining them and not merely out of regard for him.[14]

All religions require us to look upon life as an opportunity for self-realization — *ātmanastu kāmāya*. They call upon us to strive incessantly and wrest the immortal from the mortal. God is the universal reality, wisdom and love and we are His children irrespective of race or religious belief. Within each incarnate soul dwells the god-consciousness which we must seek out and awaken. When mankind awakes to the truth, universal brotherhood will follow, the at-one-ment with the great fountainhead of all creation. One whose life is rooted in the experience of the Supreme spontaneously develops love for all creation. He will be free from hatred for any man. He will not look upon human beings as though they were irresponsible things, means to other peoples' interests. He will boldly work for a society in which man can be free and fearless, a subject, not an object. He will oppose terror and cruelty and stand by the outcast and the refugee. He will give voice to those who have no voice. What gave Marxism its immense vitality is the vision of injustice made good, of the poor raised to power and the proud brought low.

Religion in this sense will be the binding force which will deepen the solidarity of human society. The encounter of the different religions has brought up the question whether they could live side by side or whether one of them would supersede the others. Mankind at each period of its history cherishes the illusion of the finality of its existing modes of knowledge. This illusion breeds

[14] Parikṣya bhikṣava grāhyam madvaco na tu gauravāt.

intolerance and fanaticism. The world has bled and suffered from the disease of dogmatism, of conformity. Those who are conscious of a mission to bring the rest of humanity to their own way of life have been aggressive towards other ways of life. This ambition to make disciples of all nations is not the invention of the Religion. If we look upon our dogmatic formulations as approximation to the truth and not truth itself, then we must be prepared to modify them if we find other propositions which enter deeper into reality. On such a view it will be illogical for us to hold that any system of theology is an official, orthodox, obligatory and final presentation of truth.[15] Reality is larger than any system of theology, however large.

All great religions preach respect for other ways of life, whatever their practices may be. It is well known that in the East religious feuds have been relatively unknown. Early Christianity was not authoritarian. It was humanistic and tolerant so long as it was the religion of the poor and humble peasants, artisans and slaves, but when it became the religion of the Roman Empire authoritarianism became more prominent. The tension between the two never ceased. It is illustrated by the conflict between Augustine and Pelagius, between the Catholic civilization and the many heretical groups and between the various sects within Protestantism. So long as this attitude persists, intolerance is inevitable. Faith without wisdom, without tolerance and respect for others' ways of life is a dangerous thing. The Crusaders who marched their armies eastward could not conceive it to be possible that the God of Islam might be

[15] Cf. Charles E. Raven: 'It is precisely this claim to an absolute finality whether in the Church or the Scriptures or in Jesus Christ or in anything else, this claim that revelation belongs to a totally different order of reality from discovery or that a creed is something more than a working hypothesis, that perplexes and affronts those of us who have a proper sense of our own limitations.'

the same God on whom they themselves relied. The historian of the Crusades, Mr. Steven Runciman, concludes his account with very significant words which have a bearing on the contemporary world situation:

> In the long sequence of interaction and fusion between Orient and Occident out of which our civilization has grown, the Crusades were a tragic and destructive episode. The historian, as he gazes back across the centuries, must find his admiration overcast by sorrow at the witness that it bears to the limitations of human nature. There was so much courage and so little honour, so much devotion and so little understanding. High ideals were besmirched by cruelty and greed, enterprise and endurance by a blind and narrow self-righteousness; and the Holy War itself was nothing more than a long act of intolerance in the name of God, which is the sin against the Holy Ghost.[16]

The *Qurān* asks us 'not to revile those whom others worship besides Allāh lest they, out of spite, revile *Allāh* in their ignorance.'[17] The *Qurān* says 'We believe in God and the revelation given to us and to Abraham, Ismail, Isaac, Jacob and the Tribes, and that given to Moses and Jesus and that given to other Apostles from their Lord. We make no difference between one and another of them, for we bow to *Allāh*.'[18] Muhammad thought of himself as one who purified the ancient faith and rid it of the extravagances that had crept into it. The *Qurān* says: 'The same religion has He established for you as that which He enjoined on Noah, which we have sent by inspiration to thee. And that which we enjoined on Abraham, Moses and Jesus, namely, that you should remain steadfast in religion and make no divisions therein.'[19]

---

[16] *A History of the Crusades,* Vol. III (1954). p. 480.

[17] VI, 108.

[18] II, 136.

[19] XIII, 13.

A religion which brings together the divine revelation in nature and history with the inner revelation in the life of the spirit can serve as the basic of the world order, as the religion of the future. Whatever point of view we start from, Hindu or Muslim, Buddhist or Christian, if we are sincere in our intention and earnest in our effort, we get to the Supreme. We are members of the one Invisible Church of God or one Fellowship of the Spirit, though we may belong to this or that visible Church.

* * *

Interreligious understanding is native to India, Aśoka in his twelfth edict proclaimed:

> He who does reverence to his own sect, while disparaging the sects of others, wholly from attachment to his own, with intent to enhance the glory of his own sect, in reality, by such conduct, inflicts the severest injury on his own sect. Concord, therefore, is meritorious, to wit, hearkening and hearkening willingly to the law of piety as accepted by other people.

Gandhi said: 'I hold that it is the duty of every cultured man or woman to read sympathetically the Scriptures of the world. A friendly study of the world's religions is a sacred duty.' We must have the richness of the various traditions. We are the heirs of the heritage of the whole of humanity and not merely of our nation or religion. This view is being increasingly stressed in western religious circles. Archbishop William Temple puts it in a different way:

> All that is noble in the non-Christian systems of thought or conduct or worship is the work of Christ upon them and within them. By the Word of God — that is to say, by

> Jesus Christ — Isaiah and Plato and Zoroaster and [the] Buddha and Confucius conceived and uttered such truth as they declared. There is only one divine light, and every man in his measure is enlightened by it. Yet, each has only a few rays of that light, which needs all the wisdom of all the human traditions to manifest the entire compass of its spectrum.[20]

Dr. Albert Schweitzer observes: 'Western and Indian philosophers must not contend in the spirit that aims at the one proving itself right in opposition to the other. Both must be moving towards a way of thinking which shall eventually be shared in common by all mankind.'[21] Professor Arnold Toynbee writes that he would 'express his personal belief that the four higher religions that were alive in the age in which he was living were four variations on a single theme and that, if all the four components of this heavenly music of the spheres could be audible on each simultaneously, and with equal clarity to one pair of human ears, the happy hearer would find himself listening not to a discord, but to a harmony.'[22] In an article in *The Observer*, October 24, 1954, he writes that 'this Catholic-minded Indian religious spirit is the way of salvation for human beings of all religions in an age in which we have to learn to live as a single family if we are not to destroy ourselves.' We should not wish any religion to compromise or capitulate. We should treat all religions as friendly partners in the supreme task of nourishing the spiritual life of mankind. When they begin to fertilize one another, they will supply the soul which this world is seeking.

---

[20] *Readings in Saint John's Gospel*, First Series (1939).

[21] George Seaver: *Albert Schweitzer* (1947), p. 276.

[22] *A Study of History*, Vol. VII (1954), p. 428.

Religion reflects both God and man. As religion is a life to be lived, not a theory to be accepted or a belief to be adhered to, it allows scope and validity to varied approaches to the Divine.

CHAPTER 2

INDIAN RELIGIOUS THOUGHT AND CIVILIZATION

The problem of religion arises from the realization of the imperfect condition of man. Life is not merely a physical phenomenon or a biological process. Who shall save me from the body of this death, from the snares and dangers of this world? The need for redemption implies the presence of conditions and circumstances from which we seek escape or liberation.

The fundamental concepts of Indian religious life may first be briefly indicated. The goal of life is communion with the Supreme. It is a life of realization, a *gnosis*, an inner intuitive vision of God, when man achieves absolute freedom and escapes from the blind servitude to ordinary experience. It is a subtle interwovenness with the realities of the spiritual world. It is not knowledge or the recognition of universal ideas through a dialectical process or analysis of empirical data. It is analogous to Plato's vision of an irresistible harmony with the deepest reality of the world inspired and sustained by the spiritual in us.

asti brahmeti ced veda parokṣam jñānam eva tat; asmi (aham) brahmeti ced veda aparokṣam tat tu kathyate.

This brings out the distinction between intellectual recognition and spiritual realization. We can free ourselves from the shackles of the body and in a split second we can see the truth and be overcome by it. We see God so intensely that the soul is more certain and more possessed by the sight of God than the bodily eye by the light of day.

tad viṣṇoḥ paramam padam
sadā paśyanti sūrayah, divīva cakṣur ātatam.

The *Bṛhadāraṇyaka Upaniṣad* tells us that through *śravṇa, manana* and *nīdidhyāsana*, we have to attain *ātma-darśana,*[1] *ātma-darśanam uddiśya vedānta śravaṇa manana nididhyāsanam kartavyam ity arthaḥ.* The *Muṇḍaka Upaniṣad* says:

praṇvo dhanuḥ śaro hy ātmā brahma tal lakṣyam ucyate:
apramattena veddhavyam saravat tanmayo bhavet.[2]

The other *Upaniṣads* say:

vedāham etam puruṣam mahāntám ādityavarnam tamasaḥ parastāt.[3]
anubhūtiṁ vinā mūḍho vṛthā brahmaṇi modate.[4]

---

[1] IV, 4-5.
[2] II, 2-4.
[3] See *Svetāśvatara Upaniṣad,* III, 8; see also III 21.
[4] *Maitreyopaniṣad,* 2.

Intution is not emotion but the claim to certain knowledge. It gives us a sense of divine reality as a thing immediately certain and directly known. The sense of God penetrates the seer's consciousness but it does not come like the light of day, something external, something out there in space. The barrier that separates the seer from the divine life is broken down. It is the aim of the seer to live in the light and inspiration of this experience, to be one with God in an abiding union.

The records of these experiences are the *Vedas*, 'ever the same yet changing ever? The *Vedas* which constitute the essential foundation of the entire spiritual tradition of India are based on integral experience. The term *Veda*, derived from the root *vid*, refers to a doctrine based not on faith or revelation but on a higher knowledge attained through a process of intuition or seeing. The *Vedas* are seen by the *ṛṣis* the seers of the earliest time. The *Vedas* do not give us theories or theologies. The hymns contain reflection of a consciousness that is in communion with metaphysical reality. The gods themselves are not mere images but projections of the experience of significance, of forces directly perceived in man, in nature, or beyond. The *Vedas* are neither infallible nor all-inclusive. Spiritual truth is a far greater thing than the scriptures. We recognize the truth and value of much that has been proclaimed by non-*Vedic* prophets and we are led equally to perceive the insight of many religious teachers in later centuries. The *Veda* is a record of inspired wisdom and deep inner experience.[5]

The second factor is the emphasis on the divine possibilities of man. The great text, *tat tvam asi,* stresses this truth. The Supreme is in the soul of man. For the *Upaniṣads*, as for Plato[6] and Philo,[7] man is a celestial plant.

[5] *tad vacanäd āmnāyasya prāmaṇyam.—Vaiśeṣika Sütra.*

[6] *Timaeus, 90.*

[7] *De plantatione*, sec. 17; cf. Seneca: 'The place which God occupies in this world is filled by the spirit in man.'

Godhead can be described and approached in various ways. The Hindu thinkers were conscious of the immensity, the infinity, the inexhaustibility and the mysteriousness of the Supreme Spirit. A negative theology develops. Brahman is a reality which transcends space and time and so it is greater than human understanding can grasp, *sānto' yam ātmā.* Brahman is silence. Yet Brahman is the continuing power which pervades and upholds the world. He is the real of the real, the foundation on which the world rests. He is essential freedom. His different functions of creation, preservation and perfection are personalized in the forms of *Brahmā, Viṣṇu* and *Siva.* The individual deities are affiliated to one or the other. When approaching the different conceptions and representations of the Supreme, the Hindu has a sense of humility, a deep awareness of human frailty. Even if religion's claim to be the results of divine revelation, the forms and contents are necessarily the products of the human mind.

> *eṣa devo viśvakarmā mahātmā sadā janānāṁ hṛdaye sanniviṣṭaḥ; hṛdā manīṣā manasābhikḷpto ya etad vidur amṛtās te bhavanti.*[8]

Religion reflects both God and man. As religion is a life to be lived, not a theory to be accepted or a belief to be adhered to, it allows scope and validity to varied approaches to the Divine. There may be different revelations of the Divine but they are all forms of the Supreme. If we surround our souls with a shell, national pride, racial superiority, frozen articles of faith and empty presumptions of castes and classes, we stifle and suppress the breath of the spirit. The *Upaniṣads* are clear that the flame is the same even though the types of fuel used may vary. Though cows are of many colours, their milk is of one colour; the truth

[8] *Śvetāśvatara Upaniṣad,* IV. 17.

is one like the milk while the forms used are many like the cows.[9] Again, the *Bhagavāta* says even as the several senses discern the different qualities of one object, so also the different scriptures indicate the many aspects of the one Supreme.[10]

In the *Upaniṣads* we find a fourfold status of the Supreme Reality—*ātmā catuṣpāt, Brahman, Iśvara, Hiraṇyagarbha* and *Virāj*. While the world is the form of the Divine, *viśvarūpa*, the cause is threefold—*pādo'sya sarvā bhūtāni tripādasyāmṛtam divi.*[11]

The problem facing man is the conflict between the divine and the undivine in him. *Yoga-sūtra Bhāṣya* says that the stream of mind flows in two directions, the one leading to virtue, the other to vice: *citta-nāḍī nāma ubhayato vāhinī, vahati-kalyāṇāya vahati ca pāpāya.*[12] To overcome this conflict and integrate the personality is the aim of religion. This problem has no meaning for beasts and gods, as Aristotle says, 'It concerns the human predicament.'[13]

There are different recognized pathways by which the duality is overcome and perfection is reached. In order to see in the world of spiritual reality, we must close our eyes to the world of nature. The *Kaṭha Upaniṣad* says that man

---

9 *gavām aneka varnānām kṣīrasyāsty eka varṇatā*
kṣīravat paśyate jñānaṁ linginas tu gavām yathā

10 *yathendriyaiḥ pṛthag dvāraiḥ artho bahu-guṇāśrayaḥ*
eko nanā īyate tadvat bhagavān śāstra-var tmabhiḥ.

11 *Ṛg Veda.*

12 I. 12.

13 *dvau eva cintayā muktau paramānande āplutau*
yo vimūdho jaḍo bālo yo guṇebhyaḥ paraṁgatḥa.
Two are free from care and steeped in bliss; the child inert and ignorant and he who goes beyond the (threefold) attributes.
Cf. Śaṁkara: *nistraiguṇye pathi vicaratām ko vidhiḥ ko niṣedhah.*

is turned outward by his senses and so loses contact with his own deepest self. His soul has become immersed in outer things, in power and possessions. It must turn round to find its right direction and find the meanings and realities it has missed.[14] To hear the melodies of spirit we must shut off the noise of the world. This is not to renounce the powers of sight, hearing and speech. It is to open the inner eye to spiritual realities, capture the sounds that come from the world of spirit, sing in silence the hymn of praise to the Supreme Being.

True religious life must express itself in love and aim at the unity of mankind. Bead necklaces, rosaries, triple paint on forehead, or putting on ashes, pilgrimages, baths in holy rivers, meditation, or image worship do not purify a man as service of fellow-creatures does.[15] The Hindu dreamed of universal peace and clothed his dreams in imperishable language:

mātā ca pārvatī devi pitā devo mahesvaraḥ.
bāndhavāḥ śivabhaktāś ca svadeśo bhuvanatrayam.
udāra-caritānām tu vasudhaiva kuṭumbakam vārāṇaśi medinī.

The goal of world unity is to be achieved by *ahiṁsā* which is insisted on by Hinduism, Buddhism and Jainism.

The fact that the Tamil classic *Tirukkural* is claimed by different religious sects indicates its catholicity. Its emphasis on *ahiṁsā* or non-violence in its varied applications — ethical, economic and social — shows the importance which ancient Tamil culture gave to it.

---

[14] II, 1,1.

[15] *rudrākṣam, tulasi-kāṣṭham tripuṇḍram bhasma-dhāraṇam; yātrāḥ snānāni homāś ca japā vā deva-darśanam, na ete punanti manujam yatha bhuta hite ratiḥ.*

Tirukkural is used by the Buddhists and the Jains, the Śaivites and the Vaiṣṇavites. It is called *podumurai* or common scripture.

The other two works of Tamil literature, *Silappathikāram* and *Manimekhalai,* exalt the virtues of chastity and renunciation.

Even Manu intended the message of India to be of universal application:

etad deśa-prasūtasaya sakāśād agrajanmanaḥ svam
savam caritram śikṣeran pṛthivyāṁ sarvamānavāḥ.

All the people of the world would learn from the leaders of this country the lessons for their behaviour.

There is a persistent misunderstanding that we look upon the world as an illusion and this view is attributed to Śaṁkara. The *Brahma Sūtra* clearly makes out that the world is non-existent, *nābhāva upalabdheḥ,* that it is not a mantal aberration, *na svapnādivat.* Of course Śaṁkara affirms that the world is not Brahman. As the manifestation of Brahman it is real only in a secondary sense; it has what is called *vyāvahārika sattā.* By no means is it to be dismissed as utterly unreal. It is different from *prātibhāsika sattā* or illusory existence. Śaṁkara makes out that the world is a progressive manifestation of the Supreme:

ekasyāpi kūṭa sthasya citta-tāratamyāt
jñānaiśvaryāṇām abhivyaktiḥ
pareṇa pareḥa bhūyasī bhavati.

There is another verse:

jagat trayam śāmbhava-nartana sthalī
naṭādhirājo'tra paraḥ śivaḥ svayam
sabhā naṭo raṅga ityvyasvasthitiḥ
svarūpataḥ śakti-yutāt prapañcitā.

The three words are but the dancing hall of God *Siva*. The king of dancers is the Supreme God himself. The audience, actors and the stage are evolved and ordered by the Lord from his own self in association with his *Śakti*.[16]

Though there was no missionary motive, no attempt to convert others to the Hindu faith, its influcence extended to other regions like Java, Bali, where we still have a Hindu colony, and other parts of the East. Greek leaders like Heliodorus became devotees of the Hindu faith. While missionary religions carry out propaganda and are interested in the increase of the number of their followers. Hindu religion was not what we call a proselytizing religion, though in its great days it had no objection to foreigners accepting the Hindu faith.

* * *

Buddhism which arose in India was an attempt to achieve a purer Hinduism. It may be called a heresy of Hinduism or a reform within Hinduism. The formative years of Buddhism were spent in the Hindu religious environment. It shares in a large measure the basic presuppositions of Hinduism. It is a product of the Hindu religious ethos. But soon it established itself as a distinctive religious tradition. It split early into two branches, though the nature of its thought and teaching is common to its different expressions. The *Hīnayāna* is the southern, Pali or *Theravāda* Buddhism; the *Mahāyāna* is the northern, mainly Sanskrit Buddhism. Both groups claim that they are

---

[16] *Soma-stava-rājā*, verse 40. Cf. also Sriharṣa: *tad eva rūpam ramaṇīyatāyāḥ kṣane kṣane yan navatām vidhatte:* That beautiful form appears fresh and different every moment,— *Naiṣadha.*

loyal to the teachings of the Buddha. The former is more monastic than the latter. *Mahāyāna* has been more sensitive to the religious yearnings of the people. While *Hīnāyana* places its emphasis on individual attainment of salvation, the *Mahāyāna* emphasizes the grace of the Divine. It is sometimes contended that the *Mahāyāna* Buddhism reveals a stage of truth greater than that which the Buddha gave to his followers in the Pali scriptures as they were not spiritually mature to receive the higher stage of truth.

The name Buddha means the Awakened One, from the root *budh,* to awaken. The Buddha is one who attained spiritual realization. He gives us a way based on clear knowledge, or awakening. Buddhism is a system of spiritual realization. So in Buddhism personal realization is the starting point. The religious experience of the Buddha is the fundamental source of the religious knowledge, of the Buddhists. *Udāna* says that he who attains final knowledge, fulfils the vow of celibacy, he is the Brāhmaṇa who has the right to declare the truth.[17]

From his experience of enlightenment, *bodhi,* the Buddha derived his doctrines. The fourfold truth, the nature of man and the character of the world, the cause of his predicament, the way by which man may rise above it and the state of enlightenment or release from subjection to time are the results of his own experience of truth. The Buddha shared with men those aspects of his experience which can be expressed in words. The state of enlightenment is beyond definition or description. The Buddha refused to speculate on the nature of transcendent reality. Each of us has to follow in the footsteps of the Buddha who blazed the path. Each individual has to attain the experience by his own individual effort. Only when the

[17] *vedāma-gu vuṣita-brahma-cariyo, dharmena sa* brahmavādam vadeyya.

individual himelf experiences enlightenment, he is said to know the truth or be enlightened. He is then freed from the shackles of earth-bound existence and becomes divine. The scriptures, the Pali *Tripiṭakas*, are the source for the knowledge of truth, since they record the Buddha's teachings. They are *Buddha-vacana*. The seekers of the past and the masters of the present attained salvation by devotion to the path revealed by the Buddha and placing their trust in him.

The Buddha stresses the possibility and need for each individual attaining the truth. *Hīnayāna* holds that the experience of enlightenment which was realized by the Buddha is attainable by other human individuals if they follow the path in his footsteps. Every individual has in him the possibility of becoming an *arhat*, who is superior to time and has conquered the world. The *Mahāyāna* adopts the ideal of *Bodhisattva* who, though he has attained release, out of concern and love for mankind, lived in the world where he may serve men by bestowing hope and guiding their steps. It preaches universal salvation. In *Hīnayāna* the founder of Buddhism is worshipped as the Divine. The other deities worshipped by men pay homage to the Buddha. He is said to be the instructor not only of men but of gods. He is to be adored as the saviour of men through the truth which he exemplified in his life. In the *Mahāyāna*, the earthly Buddha is the eternal Buddha who reveals himself in all worlds. Gautama Sākyamuni is an earthly incarnation of the Eternal Budhha who exists in countless worlds. All things are subject to him. All existences are the results of his creation. The nature of Godhead which has developed in the *Mahāyāna* is analogous to the Hindu conception. According to the doctrine of the *Trikāya*, the *Dharmakāya* or the body of Dharma is the ultimate first principle, the Divine from which all things proceed and to which they all return. It is the ultimate Godhead completely transcendent to the world. The next category of the Divine is the *Sambhogakāya*, the body of bliss or enlightenment. This answers to the personal God, who is the creator and preserver of the universe. He is the deity worshipped by man. *Nirmaṇakāya* is the manifestation of the Divine on earth. It is the divine incarnate in human life and history for the purpose of making the Divine known to man. *Mahāyāna* Buddhism has scope for the gracious saving power of the Divine. It is not merely by human effort but by divine grace that man attains salvation.

The Buddha recognizes diverse ways to reach the truth. But when the truth is attained, the way falls away. One need not insist that it is the only way to reach the truth. The Buddha gives us the parable of the raft. Any person who wishes to cross a dangerous river, having built a raft for this purpose would indeed be a fool if, when he had crossed, he were to put the raft on his shoulders and take it with him on his journey.[18] In China when the followers of Confucianism, Taoism and Buddhism meet and exalt their own religion, they conclude with the chorus: 'Religions are many, reason is one; we are all brothers.'[19] Prince Shotuko of Japan (seventh century A.D.) reconciled Shintoism, Confucianism and Buddhism:

> Shinto is the source and root of the Way, and shot up with the sky and the earth, it teaches man the Primal Way;

[18] *Majjhima Nikāya*, XXII. Cf. the *Upaniṣad*:
sāstrany abhyasya medhāvī jñāna-vijñāna tat paraḥ
palālam iva dhānvārṭhi tyajet granthān aseṣataḥ.
The wise one studies the scriptures intent on understanding their significance and (having found it) throws away the books as he who seeks the grain throws away the chaff.

[19] J. Estlin Carpenter: *The Place of Christianity in the Religions of the World*, p. 60.

> Confucianism is the branch and foliage of the Way, and bursting forth with man, it teaches him the Middle Way; Buddhism is the flower and fruit of the Way, and appearing after man's mental powers are matured, teaches him the Final Way. Hence to love one in preference to another, only shows man's selfish passion... indeed each new creed enlightens the old.[20]

According to the Buddha's Fourfold Truth, the nature of human existence is said to be of a fugitive and fragile character. This did not mean for the Buddha a world-negating creed with no concern for temporal affairs. The Buddha is not only the discoverer of truth but also its revealer to mankind. He shares with men the truth which he has attained. He shows men the way by which truth may be found. The middle path of religious realization is not only the end of religion but also the means by which truth is attained. The means of attaining the goal participates in the nature of the goal itself. The ethical means and the spiritual end cannot be separated. The end of enlightenment enters into the means. It is impossible for people who despise the world to produce the art and culture which enriches our world. Buddhism does not cause men to turn from the pursuits and endeavours of human life.

Buddhism purports to be a universal religion applicable to all mankind. In the *Mahāyāna*, not only one's personal salvation but that of all creatures is stressed. Through their infinite love for struggling humanity, the *Bodhisattvas* elect to postpone the final bliss of nirvāṇa to which they are entitled so that they may continue the unending labour of saving the souls of all, since all are destined for Buddhahood.

The Buddha entrusted to his followers the propagation of his doctrine. Under the patronage of Aśoka who became

[20] Inazo Nitobe: *Japan* (1931), p. 370.

a convert to Buddhism, repenting bitterly the carnage involved in the conquest of Kalinga, Buddhism became widespread in India. Aśoka ordered the precepts of Buddhism to be carved in stone columns and rocks. He enjoined his 'children', i.e. his people, to love one another, to be kind to animals, to respect all religions. This zealous Emperor 'beloved of the gods', *daivānam priya,* had relations with the countries of the Mediterranean and West Asia. He sent abroad missionaries to spread the Buddhist gospel. Tradition has it that his own son carried the doctrine to Ceylon. It has spread to many other lands from Afghanistan to Japan. It is a supraregional religion. In the process of its expansion Buddhism absorbed into itself the traditions and cultures of the different areas which have accepted its message. While accepting the beliefs and practices of the native peoples, it has helped to refine them.

* * *

According to Jainism, a *Tirthankara* is one who provides the ship to cross the world of *śaṁsāra*. The ship is the dharma. The *Tirthankara* is the *arhat*, the object of worship. Such a person revitalizes the dharma of the world. By destroying the four karmas, he attains the four eminent qualities of *ananta-jñāna* or infinite knowledge, *ananta-darśana* or infinite perception, *ananta-vīrya* or infinite power, *ananta-sukha* or infinite bliss. Endowed with these qualities he becomes an omniscient being who spends the rest of his life in the world for the good of mankind. When the self realizes its true nature it is freed from subjection to time or as it is said, it is released from rebirth. He becomes *siddha parameṣṭi*, the perfect being. The *siddha* is worshipped because he represents the final spiritual perfection. The *arhat*, the *siddha*, the *sangha* and the *dharma* are the four objects of supreme value worthy of adoration. Jainism emphasizes the potential divine stature of man and its teaching claims to be of universal application.

* * *

In Zoroastrianism there is a dualism, an open struggle between two forces. Ahura Mazda and Angra Mainyu are the two warring principles and in their struggle is grounded the drama of cosmic life and human history. The one is the principle of light, justice and the good; the other is the principle of darkness, injustice and evil. The battle between these two is decided by the victory of the good. Before the triumph of light over darkness is complete, the universe and mankind must pass through endless cycles of exhausting torment and untiring strife. Man in the world is confronted by the choice between the two principles. Since the conflict between the two principles is universal as to space and time, the choice which man must make is not differentiated and delimited by empirical boundary stones. As a matter of course, those who are called to be followers of Ahura Mazda form among themselves bonds of spiritual solidarity, having nothing to do with empirical relations between them, relations derived from considerations of race, political allegiance and racial groups. The doctrine is a universalist one.

The *Avestā* says:

> The souls of the faithful of both sexes in the Aryan countries, the Turanian countries, the Sarnatian countries, the Syrian countries, the Dacian countries, in all countries — all these do we venerate.[21]

---

[21] *Yast,* XIII, 143, 144.

Here we have an explicit definition of a universal religious community which supersedes all distinctions of race, caste and nationality. A believer wheresover he be found, is an object of veneration. In the Zoroastrian sense, a believer is one who, irrespective of his political allergiance and earthly origin, becomes a follower of Ahura Mazda in the pursuit of justice and peace.

Zarathustra teaches: 'And we worship the former religions of the world devoted to righteousness.'[22]

Persia, though defeated at Marathon and Salamis, exerted a powerful influence on the post-exilic Hebrew prophets and the Hellenic world. Immediately after the two great Athenian victories over the army and the navy of the Persians, a vast transformation is apparent in Hellenic religious life, due to the penetration of Indian and Zoroastrian ideas.

Professor Flinders Petrie, the great Egyptologist, in his excavation of Memphis, the capital of ancient Egypt, discovered in the Persian strata of the city, pottery, beads and figures of the Indian type. Commenting on it, he writes:

> The importance of the Indian colony in Memphis under the Persian empire lies in its bearing on its importation of Indian thought and the rise of the ascetic movement before Christ which culminated in western monachism.

Reverend Frank Knight writes:

> Monasteries or groups of ascetic devotees living together in a communal form and ordering their lives on rules laid down by Indians were established in Egypt by 340 B.C. It is in many ways probable that Greek Stoicism was not an

[22] *Yasna*, XVI, 3.

indigenous Hellenic product, but merely infiltration *via* Egypt of beliefs derived from the Buddhist priests of India.[23]

According to Plato, Socrates says:

> When the soul returning into itself reflects, it goes straight to what is pure, everlasting and impartial and like unto itself and being related to this cleaves unto it when the soul is alone and is not hindered. And then the soul rests from its mistakes and is like unto itself even as the Eternal is with whom the soul is now in touch.

This state of the soul is called 'wisdom', what we call *jñāna.* Dionysius who plays a relatively minor role in the epics of Homer now appears among the Olympian gods on the friezes of the Parthenon. Between the two dates the incursion of the Dionysius mysteries and the transformation of Greek religious life must be placed. This introduces a new mystical element into the traditional religion of the Hellenic world.

The dualism of the Zoroastrian philosophy underlies the Orphic attitude. The empirical world, the world of sense, of existence, is confused and tormented. Through music, contemplation, love, man can liberate himself from the sphere of sensory experience and earn spiritual immortality even now. Thus the religious world of the Greeks became familiar with the concept of spiritual community. The *ecclesia spiritualis* has been a historical reality throughout the centuries. Communities of men who recognize a solidarity unrelated to race, nation, blood, politcs, class, or caste, who are bound by a common belief in transcendental values and participation in divine grace sprang up.

[23] Quoted in G.S. Ghurye: *Indian Sadhus* (1953), p. 11.

Heraclitus calls every man a barbarian who heeds only the testimony of his senses to the exclusion of the spiritual harmonies which remain inaccessible to the corporeal ear. The Stoic thinkers declare that all men are brothers by an inescapable law of nature.

* * *

The Jewish *Bible* does not begin with the Jews. It starts with the story of Adam which in Hebrew means man, adam *Genesis* (V.I.) says: 'This is the book of the generations of man.' It does not speak of the Levite, the priest, or the Jew but of men. The children of earth are viewed as one family. They have one ancestor who is the father of all. Distinctions of caste and class, differentiation by blood or descent do not supersede the primary fact of human equality. 'Why was man created one?' ask the Rabbis and answer: 'In order that no man should say to another, "My father was greater than thine". '

Though the Jews are said to lay great stress on ceremonial piety, there is also stress on a different attitude to life: Man is made in the image of God. In his ultimate nature man partakes of the divine essence. The *Proverbs* describe the spirit of man as the candle of the Lord, a candle which has to be lit with a divine flame.

Though man is made in the 'image of God', 'the Fall of man' represents the lapse from the state of close relationship with God. Now, man possesses the image of God only potentially and not actually. To conform to the will of the Supreme, personal sanctification is essential. The flame of spirit must be kindled in each human soul. 'Thus saith the Lord God. I will put a new spirit within you; and I will take the stony heart out of their flesh and give them an heart of flesh.'[24] 'Create in me a clean heart,

[24] *Ezekiel,* II, 16, 19.

O God, and renew a right spirit within me.' It is the aim of the Jews to create a broken and a contrite heart, for God will not despise it.

For creating a new man and a new world, a 'turning of the soul' is essential. The soul of man is seen as 'the lamp of God, searching out all the recesses of the inward parts'. God said to Moses, according to *exodus*: 'Thou cans't not see my face, for there shall no man see me and live.' When the Covenant of God is written in the heart of man, the transcendent will become completely immanent, 'I have said, ye are gods and all of you are children of the Most High.' (*Psalms*)

The Hebrew *Bible* will not compromise with idolatry. 'Thou shalt have no other gods but me.' Tacitus says: 'The Jews condemn as impious all who, with perishable materials wrought into the human shape, form representations of the deity. That Being, they say, is above all and eternal, given neither to change nor decay.'[25] Philo quotes a letter written to Caligula by King Agrippa of Judaea in which it is said:

> O my Lord and master, Gaius, this temple has never, from the time of its original foundation till now, admitted any forms made by hands, because it has been the abode of God. Now pictures and images are only imitations of those gods who are perceptible to the outward senses; but it was not considered by our ancestors to be consistent with the reverence due to God to make any image or representation of the Invisible God.[26]

The Jews do not admit into their temple any image or representations made by hands, no visible likeness of him

[25] *Hist.*, V, 5.
[26] Quoted by Leon Roth: *Jewish Though as a Factor in Civilization* (1955), p. 25.

who is invisible Spirit. They stress the transcendence of God.

The great Commandment of the Jews is to 'love thy neighbour as thyself'. In *Leviticus* XIX, where we find a commentary on this principle, it is said:

> Let there be no hate in your heart for your brother; but you may make a protest to your neighbour so that he may be stopped from doing evil. Do not make attempts to get equal with one who has done you wrong, or keep hard feelings against the children of your people, but have love for your neighbour as for yourself. I am the Lord.

This principle applies not only to one's brothers or kinsmen or neighbours but to all. 'And if a man from another country is living in your land with you, do not make life hard for him; let him be to you as one of your countrymen and have love for him as for yourself; for you were living in a strange land, in the land of Egypt. I am the Lord your God.' Micah asks: 'What doth the Lord require of thee, but to act justly and to love mercy and to walk humbly with thy God.' Moses uttered the prayer: 'Would that all God's people were prophets.' Isaiah says: 'He shall judge between the nations and they shall beat their swords to ploughshares...Neither shall they learn war any more.' The weapons of war should be turned to the service of peace. The nations form one family and they are inter-responsible.

* * *

Christiantiy is the religion based on the life and experience of Jesus. The Cross becomes significant only when we make it our own, when we undergo crucifixion. Jesus bids us to walk the path which he trod, that we may share the union with God which he attained. 'Seek and ye shall find.' Each one must seek for himself if he is to find. The truth latent in every soul must become manifest in the awakened spiritual consciousness. It is Jesus 'risen in the hearts of men'. Then shall we be able to 'work in the newness of life'. All things are then made new. Those who raise themselves above their unregenerate condition are the god-men who are the manifestations of the new creation, the promise and pledge of the destiny in store for humanity. There is no one way by which spiritual 'rebornness' is attained. 'Marvel not that I have said unto thee, ye must be born again... The wind bloweth where it listeth and thou hearest the sound thereof, but cans't not tell whence it cometh and whither it goeth; so is everyone that is born of the Spirit.'[27] In the same spirit it is said: 'All Scripture is inspired by God and profitable for teaching, for reproof, for correction and for training in righteousness, that the man of God may be complete, equipped for every good work.'[28]

St. Paul says: 'Your body is the temple of the Holy Ghost which is in you.'[29] 'Know ye not that ye are the temple of God and that the Spirit of God dwelleth in you.'[30] 'Ye are the temple of the living God.'[31] For Origen, there is a blood-relationship between God and man. Though God is the source of our being, everlasting, transcendent, he is also close to our hearts, the universal Father in whom we live, move and have our being. 'Be ye therefore perfect even as your heavenly Father is perfect.'[32]

---

[27] *John,* III.

[28] II *Timothy,* III, 16-17

[29] I *Corinthians,* VI, 19.

[30] *Ibid,* III, 16.

[31] II *Corinthians,* VI, 16.

[32] *Matthew,* V, 48.

Paul, in his *Epistle to the Philippians,* says: 'Work out your own salvation with fear and trembling; for it is God who works in you, both to will and to do his good pleasure.'[33] 'Be assured of this as a certain truth, that, corrupt and earthly as human nature is, there is nevertheless in the soul of every man the fire, light, and love of God' (William Law). 'He who inwardly enters and intimately penetrates into himself gets above and beyond himself and truly mounts up to God.' The vital thing for us is not to hold the creed but to enter into the experience out of which it was devleoped. Man is an unfinished creation. He is left to seek and achieve completion. 'For this purpose the Son of God appeared that he might destroy the works of the devil.'[34] It is a war that shakes the whole cosmos; it is waged in the innermost soul of man. Love of God is the easiest way to reach salvation. John says: 'If a man saith, I love God, and hateth his brother, he is a liar.' This love is a new birth, being begotten of God. 'Whosoever is begotten of God doeth no sin because His seed abideth in him and he cannot sin because he is begotten of God,' says John. Love conquers the world, all its fears and anxieties. The practice of love is the natural result of awareness of God. Jesus looks upon the least of God's children as oneself. 'And all ye are brethren.'

'If any man loves the world, the love of the Father is not in *him.* For all that is in the world, the lust of the flesh, and the lust of the eyes and the vain glory of life, is not of the Father, but of the world.' We must love even our enemies. 'He that is without sin among you, let him first cast a stone.'[35]

---

[33] II, 12-13.

[34] I *John,* III, 8.

[35] *John,* VIII, 7.

The Cross means physical suffering, earthly defeat but spiritual victory. Through suffering lies the way to liberation. Pascal says that Jesus struggles with death until the end of the world. In this boundless Gethsemane which is the life of the universe, we have to struggle on unto death, wherever a tear falls, whenever a heart is seized with despair, wherever an injustice or an act of violence is committed. 'Hast thou seen thy brother? Then thou hast seen God.' This was the motto which the early Christian had, as reported by Clement of Alexandria and Tertullian. The message is of universal applicability.

> God that made the world and all thing therein... hath made of one blood all nations of men for to dwell on all the face of the earth. For in Him we live, we move and have our being; as certain also of your own poets have said, for we are all His offsprings? (St. Paul).

Existentialismm, first used by Kierkegaard in the technical sense, is the doctrine which stresses subjectivity. He holds that subjectivtiy is truth. It is a protest against Hegelianism which holds that we can reason our way to truth. The riddles of existence cannot be solved by speculative means. For Kierkegaard, truth can be found only by passionate search, by the existential commitment of the whole personality. Truth is inwardness. Kierkegaard says in his *Journals:* 'The purpose of this life is... to be brought to the highest pitch of world-weariness.'

Heidegger asks us to pass from unauthentic existence to authentic existence, from *saṁsāra* to *mokṣa* or *nirvāṇa*. For Marcel the goal is self-knowledge. It is not a problem to be solved but a mystery to be entered upon reverently.

* * *

Islam affirms that the spread of materialism brings about the downfall of great nations. The decline of the Greeks and of the Persians is ascribed to the spread of godless materialism. Theological controversies divided Christendom, and problems of social justice and brotherhood were neglected. Muhammad affirms the unity of God and the brotherhood of man. The Muslim feels deeply for the man's insignificance, the uncertainty of his fate, and the supremacy of God. Their poets, prophets and preachers enlarged on the abyss between the Creator and the creature. Though Allāh is a being without form and without parts, without beginning or end, and without equal, He must be described partially at least if He is to be apprehended by man. He is viewed as a personal being, omnipotent, omniscient, omnipresent and compassionate.

If one has to live a truly human life, *i.e.* a religious life, he must surrender his thoughts and actions to God:

> O man, Thou must strive to attain to thy Lord a hard striving until thou meet Him.
> They are losers indeed who reject the meeting of Allāh.
> They will perish indeed who call the meeting of Allāh to be a lie.
> He regulates the affairs, making clear the sign that you may be certain on meeting your Lord.

The *Qurān* says: 'Whomsoever He willeth, Allāh sendeth astray, and whomsoever he willeth He setteth on a straight path.' His transforming grace is essential for our effort to draw near to God.

The domestication of foreign elements has been in process throughout the history of Islam. While the barbarians relegated Greek thought to a few monasteries, Muslim scholars translated Greek classics, absorbed Greek thought and transmitted it later to the West where, in the twelfth century, it produced a great intellectual revival. We generally say that the European mind is made by three elements: Greek culture with its contribution of science, art and literature; Roman civilization with its code of political conduct, law and institutions; and Christiantiy. The first two are common to Islam and Christianity, and Islam believes that it has perfected and completed Christianity.

Muhammad recognized the fact that each religious teacher has faith in his own mission, and his vision and experience alone would help fulfil the needs of his people:

> There is not a people but a warner has gone among them.
> And every nation had a messenger,
> And every nation had a guide,
> And certainly We raised in every nation a messenger, saying 'Serve Allāh and shun the devil.'
> To every nation we appointed acts of devotion which they observe.
> For every one of you did We appoint a Law and a way.[36]

* * *

If there are similarities in the religious experience of mankind, it only means that a common humanity reacts in more or less similar ways to man's encounter with the Divine. The common points to be found in the different manifestations of religion should not lead us to think that they are organized in each religion in the same way. The manner in which these beliefs are correlated varies from one religion to another. Each religion is a living organization of doctrine, worship and practice, has a uniqueness and individuality of its own and changes as a

[36] *Qurān,* XXXV, 25; XVI, 37.

whole in response to the needs of the age. While therefore we indicate the area of agreement, the distinctive arrangement of the basic presuppositions gives the quality to different religions. For our present purpose, it is not necessary to stress the differences which are important and fundamental in some points. Even though each sect of a religion claims to be the true representative of its specific religious message, yet all the followers of all the sects feel that they are bound together in a unity. As we are trying to overcome the conflict within each religion where every organized group claims to possess the truth by the recognition of the unity of religion, even so conflicts among religions require to be reconciled, if religion itself is not to be defeated.

The world has bled and suffered from the disease of dogmatism, of conformity, of intolerance. People conscious of a mission to bring humanity to their own way of life, whether in religion or politics, have been aggressive towards other ways of life. The crusading spirit has spoiled the records of religions.

In future there can be only one civilization in the world, for it is no more possible for different civilizations to live in ignorance of one another. The scientific discoveries which have penetrated in all parts of the earth are making the world one, though the different civilizations live by and cherish their distinctive principles of life.

If the world is to be united on a religious basis, it will not unite on the basis of this or that religion but by a co-operation among the different religions of the world. If the different religions strive to achieve their common ideals and seek to understand the differences in a sympathetic spirit, the world will be relieved of the misery and fear which now engulf it. The tradition of opposition to one another should yield to co-operation. The conviction of superiority, which is natural, should not prevent appreciation of other faiths and fruitful interchange among them. Erasmus delivered the great dictum: 'Wherever you encounter truth, look upon it as Christianity.' We must remember the spirit of this advice when we are wandering in the obscurity of the future. If the message of religions is to be articulated in relation to the problems of our age, we must give up the view that any one religion contains the final, absolute and whole truth, and adopt the Eastern attitude that the faith is realized in historical patterns, though no one of these patterns should regard itself as the sole and exclusive truth for all.

We must be on our guard against the enemies of truth, men of fixed ideas and fanaticisms.

Between the believers in the different historical patterns, there exists a hidden common substratum. If we overlook this, we will not be able to overcome nihilism, lack of faith and irreligion.

If we seek for a joyous reconciliation of the members of the human family, we will discern that even heretics have divined some aspect of Godhead. Just as God lets his sun shine on good and evil, He pours forth His loving kindness on all the children of mankind. The witness of the different major religions strengthens the view that religion is the hope of man and can sustain the new world.

bahu-dvārasya dharmasya nehāsti viphalā kriyā.[37]

Religion has many doors; the observance of its duties can never be useless. This view makes for the appreciation of religious knowledge, of the beliefs and practices of other peoples. This understanding makes for spiritual fellowship. Within this fellowship, each religion will have scope for full expression. Religious reflection will be

[37] *Mahābhārata, Sāntiparva,* 174, 2.

stimulated by the knowledge and friendship of others of different religions. We will also have universal ethical standards. Even as the interplay of Jewish, Christian and Muslim in the West has enriched the experience of the West, that of Hindu, Buddhist and Confucian has enriched the experience in the East, so the cross fertilization of ideas among the living faiths of the world will tend to foster and enrich spiritual life. The sign of hope is the perpetual youth of religions, the way in which they renew themselves as the world changes.

* * *

Arnold Toynbee says:

> As I have gone on, Religion has come to take a more and more prominent place, till in the end it stands in the centre of the picture.... I have come back to a belief that Religion holds the key to the mystery of existence; but I have not come back to the belief that this key is in the hands of my ancestral Religion exclusively... The Indian religions are not exclusive-minded. They are ready to allow that there may be alternative approaches to the mystery. I feel sure that in this they are right, and that this catholic-minded Indian religious spirit is the way of salvation for all religions in an age in which we have to learn to live as a single family if we are not to destroy ourselves.

The choice before humanity is, either co-operation in a spirit of freedom and understanding or conflict in an atmosphere of fear, suspicion and jealousy. The future of religion and mankind will depend on the choice we make. Concord, not discord, will contribute spiritual values to the life of mankind. Concord alone is meritorious, said Aśoka: *Samavāya eva sādhuḥ.*

Religion has emerged mature from the criticism of science and social conscience; when faiths interact, our own religion is imperceptibly modified. The unreal yields to the real.

CHAPTER 3

THE WORLD COMMUNITIES OF IDEALS

If there is any phenomenon which is characteristic of our times, it is the mingling of peoples, races, cultures and religions. Never before has such a meeting taken place in the history of our world.

Civilizations went on in parallel lines, remote from one another, unaffected by one another, but today that is not possible. The inventions of science and technology, the political concepts, and the economic ideas are bringing the world into a close neighbourhood, and it is our great hope that this neighbourhood, may be transformed into a true brotherhood. The world must become our home, if we are to save the human race.

On the one side, we see many indications which give us hope and assurance: the United Nations Organisation, the ILO, UNESCO, WHO, and others. There are also many obstacles to world unity.

Take the spaceships. In ordinary times, they would have been welcomed as a great demonstration of man's intellectual penetration. We would have welcomed them for the possibilities which they contained of conquering space; we would have utilized all these scientific powers for the advancement of human welfare. But the world was not very happy when these space vehicles travelled around. We were afraid as to how they were likely to be employed in this divided world.

Will the divided world degenerate into a world that will be destroyed? Or will it lead to a world which will be unified? Is it the beginning or the end of a new era? Is it the prologue or the epilogue? There are explosive forces all round, especially in Asia and Africa. We hear protests against race discrimination. We have underprivileged countries which are seeking to better their conditions.

The world is expected to be our home, yet the world is seething with explosive forces. What are we to do? Is it something which is beyond our range, to bring the world into a close unity, based on community of ideals? Is it difficult?

What is the obstacle that stands in the way of achieving the prophets' dream, one world, peace and goodwill on earth? What stands in the way is not lack of material resources, is not lack of intellectual power or skill, but a kind of cussedness in human nature, greed, vanity, prestige, honour, etc.

The new world is a call to us that we have to readjust ourselves to it. The old social and economic patterns will not do. We have to change, fashion a new type of human being who is relevant to the new world in which we happen to be.

This task of refashioning the individual, remaking him, is the task generally assigned to the discipline of religion.

It is that discipline which asks us to look within ourselves and to transform ourselves, to cleanse ourselves of all evil tendencies, the baser, the fallen side of human nature, and raise ourselves to a higher plane.

Unfortunately, today, religions themselves are passing through a mood of criticism. We cannot accept in this scientific age incredible dogmas, doubtful events; we want to have a religious faith which commends itself to man's understanding, to the spirit of reason.

Again, we have the other difficulty that religious leaders somehow do not rise to the occasion when great injustices occur. It is the duty of the religious leaders to stand above national politics, to urge the people to change themselves. Religious leaders have not been able to rise to the occasion. Either we said religions and social order were divorced from each other, 'Render unto Caesar', or the penetration of the world has been so intense that religions got adjusted to the world itself. That is a thing which critics of religion urge against us.

Besides, if we want to achieve world unity, religion must have a universality of outlook. But it has become like the Nation State, a bad citizen belongs to it, a good stranger is out of it, an alien. We are adopting more or less the same policy, even with regard to religions themselves.

* * *

These three formidable obstacles, the spirit of science and criticism, the awakened social conscience which protests against the inequities which are being practised in the world, and the provincialism of religions which, instead of helping one another, are competing with one another, are making intelligent people doubt the value and validity of religion. Then there is the increasing secularization of the world.

Professor C.S. Lewis, in his inaugural address which he gave at Cambridge in 1954, made out that the history of Europe had three periods, the Pre-Christian, the Christian and the Post-Christian. He meant that the Europeans were pagans once, then they became Christians, but a process of de-Christianization has started. It started somewhere about the end of the seventeenth century. Now we want to check it, we want to restore the place of spiritual values in human life. That is our great concern. And if we wish to do that, we have to reckon with the great challenges which are confronting religions.

I am sure that religions which are now passing through a process of self-understanding, self-searching, self-criticism, will be able to respond adequately to these great challenges. I am a firm believer in the need for religion and the need for co-operation among religions. I feel that there is no opposition between religion and science, between religion and highest social morality, between religion and co-operation among religions.

If we take up the scientific attitude, what is it we find? The scientists look at the world and are able to observe that this world has been an ordered one, has been a progressing one, it has grown from a state of mere materiality to one of life, from life to animal consciousness, from animal consciousness to human intelligence. It has to grow from the level of human intelligence to that of spirituality. Cosmic evolution has not come to a stop with the advent of intelligence. The further evolution will not be in the physical make up of the man; it will be in his psychical nature.

It is that psyche of the human individual that has to grow, that has to expand. The purpose of religion is to help us to grow from this world of intellect, this world of divided consciousness, with its discords, dualities, to a life of harmony, of freedom, of love.

Hitherto, in the sub-human species, the progress took place automatically. But at the human level, man has to put forth effort to achieve his goal. He is no more a mere spectator; he is a participant in this process of cosmic evolution.

It is wrong for us to think that we are the victims of natural forces, that there is a kind of inevitability, that inexorable laws prevail, that man cannot help; he has only to endure whatever happens. Man is intended for something greater than confinement in this world. He can rise above it. He can defy nature.

So if we are to rise from a state of intellectuality to spirituality, it is an effort which we have to make, and it is an effort which we can make, because others have struggled, have striven, have achieved the goal, and what is possible for some is possible for all.

Religion, as an inward transformation, as a spiritual change, as the overcoming of the discords within our own nature — that has been the fundamental feature of it from the beginning of history.

* * *

In our country we say that we should transform our nature, grow from the slavish, unregenerate condition of ignorance to a state of wisdom. The growth or the transition from the one to the other constitutes the goal of the religious quest. From the disruption of being we must rise to the articulation of being. The Buddha said exactly the same thing. We are sunk in suffering and ignorance, and our goal is to grow into enlightenment. According to Ezekiel: 'Thus saith the Lord God — I will put a new spirit within you; and I will take the stony heart out of their flesh and will give them an heart of flesh.' For the Jews, 'the spirit of man is the candle of the Lord.'[1] Speaking of the mystery religions of Greece, Aristotle observes: 'The initiated do not learn anything so much as feel certain emotions and are put in a certain frame of mind.'

To live one must first die to his old life. Orpheus believed that the soul was 'the son of the starry heaven,' that its dwelling in a body is a form of original sin, its earthly life was a source of corruption and its natural aim was to transcend this life. This view is at the heart of Plato's idealism. Plato gives us in his image of the Cave in the Republic, that we are all prisoners living in shadows. One philosopher turned round, and freed himself from his shackles. When the philosopher left the cave he saw the sun shining of which the fire in the cave was a small reflection. After having seen the great light Plato's philosopher does not remain content with his own revelation. He returns to the cave and talks to the prisoners shackled there that what they take for reality is only a shadow cast by the light they do not see. The prisoners not having seen the light take the shadows to be the only reality and think that the philosopher is insane.

Philosophy, for Plato, is the love of wisdom, the fine flower of serenity, Plotinus says:

> Withdraw into yourself and look, and if you do not find yourself beautiful yet, act as does the creator of a statue that is to be made beautiful: he cuts away here, he smoothes there, he makes this line lighter, this other purer, until a lovely face has grown upon his work. So do you also; cut away all that is excessive, straighten all that is crooked, bring light to all that is overcast, !abour to make all one glow of beauty, and never cease chiselling your statue, until there shall shine out on you from it the godlike splendour of virtue, until you shall see the perfect goodness surely established in the stainless shrine.

---

[1] *Proverbs,* XX. 27.

The author of the *Fourth Gospel* makes Jesus say: 'I am the Truth.' The religion of truth is based on spiritual inwardness. The descent of the spirit at Jesus's baptism or of his temptation in the wilderness must have been the story of his inner experience. In their present form they are externalized.

In Christianity we are called upon to follow the example of Jesus. We are to be made like unto him by bringing our natural desires and expectations into subjection to the Universal Purpose. William Law says: 'To have salvation from Christ is nothing else but to be made like unto him; it is to have his humility and meekness, his love of God, his desire of doing God's will.' Jesus asks us to free ourselves from priestly control, and undergo spiritual growth. We must be born again, born of the spirit of Truth. A Sufi mystic of twelfth century, Ayn at-qudāt al Hamadhāni (A.D. 1131) says: 'He who is born from the womb sees only this world, only he who is born out of himself sees the other world.' Ibn' Arabi of thirteenth century says: 'I am knowledge, the known and the knower. I am wisdom, the wise man and his wiseness.' (60.16). Both the Buddha and Jesus tell us: 'Be of good courage. I have overcome the world.'

Religion is spiritual change, an inward transformation. It is a transition from darkness to light, from an unregenerate to a regenerate condition. It is an awakening, a rebornness. We must break the bonds that are laid on us by our first birth and rise above our original imperfection through blood and tears.

By groaning and travailing we rise from division and conflict into freedom and love. The flame of the spirit has to be kindled in each individual soul. This is not the result of the acceptance of dogmas or historic events. We must get across the frontiers of formulas and the rigidities of regulations. Religion is an experience which affects our entire being, ends our disquiet and anguish, the sense of aimlessness of our fragile and fugitive existence. This state may appear to be one of retreat, of escape from a threatening world. The mystic claims that the realization of his yearning is far richer and deeper than the deepest satisfactions of this world.

It can be said that man, when he feels lonely, inadequate and incomplete, in the shock of his loneliness or isolation craves union with the Ultimate Reality. When he has this contact he gets back to the world and loves and serves his fellow-men spontaneously. The cosmic process has for its goal the kingdom of free spirit where the son of man becomes the son of God. The first fruits of the new species of spiritual personality are already manifest on earth in the saints and sages of the different religions.

* * *

No language is adequate to describe the existential character of the spiritual experience, the ravishment of the soul when it meets in its own depths the ground of its life and reality. This is the ultimate religious evidence. Whatever their race, creed or nationality may be, the evidence of these seers is unanimous. William Law says, that it is a cause of profound thankfulness that 'so many eminent spirits, partakers of the divine life have appeared in so many parts of the heathen world, glorious names; sons of wisdom, Apostles of a Christ within.' These spirits who hold that religion is not an academic discipline, but a mystery to be lived have more in common with each other than with the bulk of the adherents of their own religions. The destinies of the new religion are bound up with their views than with the findings of priests and scholars.

In theories of religion, the being of the soul is made into an object. All religions are human attempts to reach the Ultimate Reality. The great mystic philosopher, Nicholas of Cusa, penetrated to the discovery that 'God is sought in various ways and called by various names in the various religions, that he has sent various prophets and teachers in various ages to the various peoples.' It is one of the tragic confusions of religious history that as a faith becomes credal, the creed by which man communes with the Divine supplants the Divine. The prophet who announces the message becomes himself an object of worship supplanting the higher truth in which he believes. We become ambitious for our formulas, for our prophets, for our organizations.

The menace to religion as spiritual adventure is the claim of final solution. A regimented mind is not suited for spiritual purposes. We should recognize alternative approaches to the mystery of God. We all seek the same goal, though under different banners. Each one's life is a road to himself to self-realization.

It is difficult for us to adopt today the view that the Scriptures are literally inspired, that every word of them should be treated as factually true. Intellectual authority is superior to inherited authority. Scriptures are the records of the experiences of the great seers who have expressed their sense of the inner meaning of the world through their intense thought and deep imagination. Scientific theories and historical statements cannot be integral parts of religious Scriptures.

The seers, to whatever religion they may belong, agree that man is confronted by something greater than himself which, in contrast to human nature and all other phenomena, is Absolute Reality. It is also Absolute Good for which man is athirst, that needs not only to be aware of it; but to be in touch with it. That is the condition in which he finds himself at home in the world. After enlightenment Fox found that 'all creation gave another smell beyond what words can utter.' When we have the experience of the Reality, we try to preserve its memory by attempting to convey it through words. We know that no finite form can convey adequately the sense of infinitude.

The Greeks, the Indians and the Chinese do not look upon religious theories as giving literal interpretations of the experience. They are symbolical representatives of the intuited truth. The Supreme is above all religious systems. He is without end or limit though theologians attempt to set limits to the Supreme. There are historical views which are sometimes alien to the essential truths. These are accepted for communicating the message to the members of a society in a particular stage of history. What is permanent and universal is translated into something temporary and local.

The *Qurān* states that every nation has been given divine guidance. 'Thou (O Muhammad) art but one who warneth, and for every nation there is a (divine) guide.'[2] And again: 'Unto every people did we send a messenger, to teach them to worship God.'[3] There are as many ways to God as there are souls whom he has made. Each person is unique and his relation to God is also unique.

* * *

When religion becomes organized, man ceases to be free. If we think that it is a question of life or death — what concept of God we accept, then our hearts are filled with fury. It is not God that is worshipped but the group or the

[2] Sura 18-5-37.
[3] Sura 16-5-37.

authority that claims to speak in his name. Sin becomes disobedience to authority and not violation of integrity. For Simone Weil (1909-1943) faith is to believe that God is love and nothing else. Everything else including theological dogmas and 'the unconditional and global adherence to all that the Church has taught and will teach, which St. Thomas calls faith, is not faith but social idolatry.' Arguments in favour of official Christianity, she said, 'sound like the slogans for "Pink pills for Pale people".' To look for shelter or consolation in religion is error. 'Religion as a source of consolation and peace of mind becomes a sort of an advertised patent medicine.' As such 'it is an obstacle to true faith. In this sense atheism is purification.'[4]

Goethe's drama *Faust* begins with a Heavenly prologue in which God and the Devil discuss the highest image of the Deity which man can frame when worshipping him. But the God and the Devil agree that the God as imagined by man is a pitiable creature when compared with the Reality of God. We should not confuse the image of God as man conceives Him, with the Reality of God. The different images are to be accepted as aspects of a deeper unity. We should perceive spiritual unity beneath the divergent symbols and individual preferences.

We should not look upon our religious heritage as an invisible whole. We should make a distinction between the spirit of religion and the forms, ceremonial, ritual, marriage customs, food rules and social organization which are its forms. Accidental accretions are not as valid as spiritual truths.

Religion is not compatible with moral ease. Men cannot stand away from social order. We must love our neighbour

[4] Quoted in Sorokin: *The Ways and Powers of Love* (1954), p. 163.

in an effective sense. Though your Constitution says that all men are born equal, it means that though we are not equal in many respects, we are equal in this that we are called upon to face suffering and endure it. We must, therefore, develop compassion. The path of spiritual growth leads away from egoism, self-will, from a sense of personal superiority, in the direction of humility, openness of mind, a sympathetic understanding of the needs of others and a willingness to take responsibility. Talmud has it: 'Would that they have forgotten my name and done that which I commanded of them.' In international relations also, we should adopt an attitude of forbearance and understanding. They are invincible. The important point about our moral life is, not whether we are Hindus or Muslims, Jews or Christians, but whether we are good or bad.

The mingling of peoples is compelling us to define our attitude toward faiths other than our own. We are obliged to look at our religions in the light of other faiths. We should see them with new eyes. Since you are familiar with Christianity, I may use its developments to illustrate the variety of views held in regard to the relations to other faiths.

For Karl Barth, non-Christian religions are invitations of the devil to draw men away from the truth. He says: 'God's revelation is the annulment of religion', taking religion to mean a system of thought and culture. He was repelled by the liberal theologians of the late nineteenth century who sought natural and rational explanations of the supernatural events recorded in the Christian Scripture. They looked upon the Gospels as the uncertain and fragmentary recollections of the impact of a great prophet on his contemporaries. It is also said that they are the poetic expressions of the great truths of religion. The truth of Christianity for Karl Barth is that Jesus behaved as God and man and he was capable of suspending the laws of nature which derive from the will of God. So he makes out that religions framed by men are mere self-assertions, forms of unbelief; attempts at self-justification and self-redemption. For him, even Christianity as a religion is one among others. Self-disclosure in Christ is the fulfilment of man's needs and is a judgment on all man-made religions. Christ stands as fulfiller and judge of contemporary and empirical Christianity, as of other religions. Both Christian and non-Christian religions are condemned as sacrilegious human attempts at self-justification in contrast to the Biblical revelation of God. Barth emphasizes the sovereignty of God. All have sinned and fall short of the glory of God. Apparently, he overlooks the other saying: 'The Creation waits with eager longing for the sons of God to be revealed.'

The history of religions illustrates the tragic effect of an intolerant and exclusive faith. If we adopt Barth's view and look upon God as a jealous one there will be no peace in the religious world. This view of religion has been a stumbling block to sensitive souls and led to the abandonment of religion by them. I am persuaded that this view of Christianity which led to the Inquisition and the Wars of Religions is not fair to the teaching of Jesus that God is love.[5] To think that any human being or institution has the monopoly of God's truth is to commit the sin of pride. 'Think not to say within yourselves, we have Abraham to our father for God is able of these stones to

[5] Gandhi wrote to an American Missionary who claimed that the Christian way is the best for all: 'You assume knowledge of all people which you can do only if you were God. I want you to understand that you are labouring under a double fallacy. That what you think best for you is really so; and that what you regard as the best for you is the best for the whole world. It is an assumption of omniscience and infallibility. I plead for a little humility.'

raise up children unto Abraham.' When we think that we possess the truth, it is inevitable that we should be hard on those who do not share it. At the root of all faiths is God who is neither Hindu nor Christian, neither Jew nor Muslim.

Others who follow Karl Barth affirm God's unique, final, full, unsurpassable revelation of Himself in Jesus Christ, and that this revelation involves a break with the past.

* * *

There is the view of universal revelation which has the support of Justin, Clement and Origen, that the Logos or Word of God inspired all that is true and good in the religious thinking of men, preparing them for God's supreme, unique, revelation in Christ. The seeds of Logos, *Logos Spermatikos* were scattered in all mankind. Justin proclaims: 'All who have lived according to the Logos are Christians, even if they are generally accounted as atheists, like Socrates and Heraclitus among the Greeks.'[6] Clement of Alexandria looked upon Greek philosophy as 'a preparation for Christ,' 'a schoolmaster to lead us to Christ.'[7] He brought about the marriage between Platonism and Christianity. The early Fathers enriched Christian mysteries by using the ideas of Socrates and Plato. Augustine's views are well-known. 'The salvation brought by the Christian religion has never been unavailable for any who was worthy of it.'[8] What is now called the Christian religion always existed in antiquity and was never absent from the beginning of the human race until Christ appeared in the flesh. 'At this time, the true religion which was already there, began to be called

[6] I, *Apology,* 46.

[7] *Stromata,* I. V. 28, 32.

[8] *Epistle C,* II. 5.

Christianity.'[9] It is now admitted that in the course of its development, Christianity has drawn upon Greek metaphysics and mystery religions. Even the religion of the New Testament, in the words of St. Paul, is 'debtor both to Greeks and to Barbarians'.[10] It is obvious that Christianity is an organic part of world religious development. It has grown, like every other religion, in a long, historical process. It did not come into the world as a ready-made supernatural system.

The other religions also bring us into contact with the eternal word of God and are sometimes called a preparation for the New Testament. Christian faith is viewed as a fulfilment of other religions.

Thomas Aquinas distinguished between General and Special Revelation. The former is common to all mankind by which men attain to the knowledge and existence and unity of God. Revealed religion is above reason though not opposed to it. It is the way of faith by which men accept the Doctrine of the Trinity and the Person of Christ as true God and true man.[11] Non-Christian religions are the result of a general revelation open to men as men, and the Christian faith is the result of a special revelation in Jesus Christ.

If God is love, it cannot be that mankind lived for thousands of years without the revelation which he gave to the tribe of Israel and the adherents of other religions were shut away from his love. When Francis Xavier went to Japan and preached to its people that God in his mercy sent his son into the world, he was asked why God waited so long before acquainting the Japanese and other people with his great love. The revelation of a God of love must embrace all nations, ages and religions. But if he restricts

---

[9] *Retractions,* I, XIII. 3.

[10] *Romans,* I. XIII. 3.

[11] See *Romans,* I. 20.

his revelation to the chosen people of the Old and the New Testaments and allows a large part of humanity to sit in darkness and death, he cannot be a God of love. God manifests himself throughout history.

If Christian religion is to be true to its main tradition, it should admit the operation of divine revelation in non-Christian religions. As knowledge of non-Christian religions is spreading in the West, the conception of the unity of all religions is slowly gaining acceptance. Apart from miracle stories, cult symbols, eschatological ideals and ecclesiastical institutions which seem to be similar in different religions, the deeper aspects are also profoundly akin.[12]

More than a hundred years ago Joseph Gorres gave impressive utterance to this fact. *One* Godhead alone is at work in the Universe, *one* religion alone prevails in it, *one* worship, *one* fundamental natural order, *one* law and *one Bible* in all. All prophets are *one* prophet; they have spoken on *one* common ground in *one* language, though in different dialects. As the great civilizations of every kind are the same, the unfolding of *one* life, so are also the great mythical elements of the whole world the same and the whole religious genesis a single growth, planted by the very Spirit of God, and nourished by him with the airs of

---

[12] Professor Friedrich Heiler in an important article on *Christian and Non-Christian Religions* writes: 'The doctrines of the Trinity and Incarnation as well as the Virgin Birth, belief in the Divine Sacrifice of love, the conception of irresistible Grace and justification by faith alone, prayer prompted by the grace of God, petition for forgiveness of sins, all-embracing love towards every creature, heroic love of enemies, belief in everlasting life, in the judgment and restoration of the world — there is not a single central doctrine of Christianity which does not have an array of striking parallels in the various non-Christian faiths.' — *Hibbert Journal, January, 1954*

heaven and the dews of earth, unfolding itself in joy throughout all ages.[13]

Even the conviction that one's own faith gives a deeper insight into reality, need not engender hostility to those who cherish other beliefs. We may look upon others as fellow-seekers of truth.

Every religion is passing through self-analysis and self-criticism and is developing into a form which is sympathetic to other religions. No religion has yet uttered its last word. No religion can retreat from modernity and science. With the spread of scientific knowledge religions are becoming liberal, though a few cling to dogma as their only defence in this predicament. They are attempting to reach to what is of fundamental importance, the common root in the spiritual world from which each individual may gain a clear insight and a firm faith in his own religion.

The goal of the universe is a deep fellowship of the spirit. All religions which today are in a process of self-understanding and spiritual exchange are getting near each other. No one need give up one's own religion and engage in a syncretism. We can learn from other religions in a spirit of mutual respect. Nothing true should be alien to us. St. Ambrose's saying which was adopted by Thomas Aquinas is worthy of acceptance. 'Every truth by whomsoever it is spoken, is of the Holy Spirit.'[14] When Francis of Assisi picked up a paper and was told that what was written on it came from a pagan writer, he replied: 'That means nothing, for all that is said, whether by pagans or anyone else, comes from the wisdom of God and has reference to God, from whom comes every good thing.'

---

[13] Quoted by Professor Friedrich Heiler in the *Hibbert Journal* (January 1954), p. 111.

[14] Thomas a Celano: *Vita* II, ed, Alencon, pp. 173 ff.

Religion has emerged mature from the criticism of science and social conscience, accepting whatever is valid in other religions. When the faiths interact, our own religion is imperceptibly modified. The unreal yields to the real. We give up the notions of chosen peoples, chosen nations and chosen creeds. If we are to create a spiritual unity which will transcend and sustain the material unity of the new world order, we need inter-religious understanding. The new religious situation will be not an endless homogeneity but an organic unity where we will have sympathetic understanding and appreciation of other faiths. All religions will express themselves as forms of the universal religion of knowledge and love and from this standpoint we will be able to criticize the past history and present doctrines of every religion with severity as well as sympathy.

Even though we follow different roads, our goal is the same, reaching the ultimate mystery. We are all engaged in the same quest. We must treat one another as spiritual brethren. Toleration should be transformed into love.

* * *

There is a movement towards unification in all religions. Disunity started fairly early in the Christian Church. Two important breaches of Christian unity occurred in the eleventh and the sixteenth centuries respectively. In 1054, the Christian Church was split into two groups, one covering the countries of Eastern Europe and Egypt and Syria with Constantinople as its main centre, the other the Catholic Church covered the countries of Western Europe with its centre in Rome. Since then the Eastern Church had remained separate. Five hundred years later in the sixteenth century, the Reformation disputed the claims of the Pope and Protestant Churches arose in Germany, France, Netherlands, Switzerland, England and Scotland and the Scandinavian countries.

Apart from the main divisions of the Catholic, the Protestant and the Eastern Orthodox Church, under each of them there are widely divergent groups. While some of the Protestants treat the essence of Christianity as the acceptance of the apocalyptic-eschatological world-view, others reject it as being the expression of the mind of the period in which those passages of the Old and New Testaments were written. For the Fundamentalist these beliefs are essential, while for the Modernist they are not. From the time of St. Paul's letter to the Church at Corinth, attempts to end the divisions have been made continuously to bring the Christian Churches of East and West together but the Roman Catholic Church does not participate in them since it is convinced that the way to Christian unity is submission to the Pope at Rome. Co-operation among the churches in matters of doctrine and common action in grappling with the problems of social life have been the main objectives of these movements.

The second Assembly of the World Council of Churches was held at Evanston. It is our hope that this movement will be extended to the living faiths of mankind. A world civilization can grow on the basis of co-operation among religions. It will broaden our vision of divine activity in life and free us from narrowness and dogmatism. All the religions are our inheritance and we should not squander it away.

There are bound to be religious differences in the world. When we wish to grow in partnership, we do not advocate an undifferentiated universalism or an easy indifferentism. We accept differences and plead for a healthy growth of unity. It is not our desire to obliterate the differences but we wish to use the differences to strengthen and enrich partnership. We must develop the right temper of mind, a world loyalty through a spirit of fellowship among mankind. We should promote free co-operation among all who believe in God or an ultimate Spiritual Reality. Whatever our religious views may be, we are all one family under God. Joachim of Fiore spoke of the coming 'Church of the Spirit'. All those who observe the two commandments, 'Thou shalt love the Lord thy God with all thy heart and with all thy soul and with all thy strength and thou shalt love thy neighbour as thyself,' belong to that universal Church.

According to the writer of the last book of the *Bible*, there will be no temple in the heavenly Jerusalem, for God will be all in all.

If we do not take note of the currents of thought and aspiration but claim to speak of infallible truth about God, discard the canons of social justice, overlook that God finds something of himself in each religion and not fully in any, if we do not develop community of minds in a world that is desperately threatened by instruments which we ourselves have devised, the number of unbelievers will increase and God himself may join the camp of unbelievers.

We believe in creative evolution. The only absolute in the world is change. We need today the courage to change, to scrap old prejudices. 'For it is not known what Man maybe.'

CHAPTER 4

INTERNATIONALISM: A Challenge & An Oppurtunity

Two prominent features of our time are the increasing unification of the world and the development of nuclear weapons. Mankind, divided into groups determined by geographical conditions and historical forces, is getting together into a single neighbourhood. Individuals who looked upon themselves as members of groups representing special interests, racial, religious and national, are becoming increasingly aware that they belong to a whole, a single family. Ties of race and history, common beliefs and loyalties which bind men together have been loosened by the forces of modern life and the strains of economic change. The inventions of science and technology which diminish distances, economic penetration, and the circulation of ideas through new means of communications are at work. Civilizations are no more remote from each other. In this shrinking world we cannot live without others. This is a condition arising out of a series of historical facts and consequences. The many experiments in international co-operation, in education, health, science, culture, labour, food and agriculture; the increasing awareness of common interests among the peoples of the world, the growing interdependence of nations indicate that we are moving towards an international society, a family of mankind. George Washington prophesied that all nations will become conscious units in 'the great Republic of Humanity at large.' The world is our home.

Yet, it is in ferment. The forces of science and technology which have helped the unification of the world are making the use of force outmoded, a relic of the past. There was a time when Governments believed in organized warfare. Even when they were defeated they did not lose faith in violence. They traced their defeat to inadequacy in military strength and tried to increase it. Today a war with modern nuclear weapons may mean the destruction of civilization. In such a conflict there will be no distinction between the victor and the vanquished. Even non-belligerent nations who do not have any nuclear weapons will be involved in it. Nuclear tests, scientists warn us, endanger the life and health of generations yet unborn. The understanding of common danger, the possibility of total annihilation, is a constant reminder to us that we should either disarm or suffer disaster.

There is no nation in the world which has not the great desire to survive and yet the great powers are increasing steadily their piles of deadly weapons, as if driven by some fatality. The unbridled race for armaments and the mounting war psychosis show that we are confused in our minds about peace and friendship among nations. Is it impossible for us to say that we are too proud to fight with nuclear weapons? It, of course, requires an act of faith. Are we incapable of it?

Though we define man as a reasoning animal, his conduct is not guided by logic and reason. Passions, vanity, honour, and short-term interests often govern national behaviour. We are victims of local pressures and national obsessions. Successive waves of internationalism broke on the shores of local pride and national vanity.

This generation has the responsibility for deciding whether the human race is to prosper or decline, whether our conduct will lead to a beginning or an end. It is a challenge that faces us with danger and opportunity, world destruction or world peace. It is the crisis of the human race. Deuteronomy tells us: 'I call heaven and earth to record this day against you, that I have set before you life and death, blessing and cursing; therefore choose life, that both thou and thy seed may live.'[1]

We are asked to choose. But man is tending to lose his creativity when he feels and nothing gets better, he cannot improve anything, his daily life is controlled by pressures and he hears from philosophers echoes of his anxiety and despair. The changeless values of spirit, the pursuit of truth and the practice of love which have nourished the great pages of history do not seem to be relevant to an age steeped in science, deserted by religion and deprived of even humanist ideals. We seem to have reached the height of irresponsibility.

It is wrong to assume that events are sweeping mankind towards unknown and predestined ends. We are not the helpless pawns of natural forces. There are no inexorable laws of historic development. If we assume historical inevitability, our effort will be paralysed, our initiative destroyed and our dignity lost. Though man is rooted in

---

[1] XXX, 10.

nature, he also transcends nature. Marx was right in holding that we are conditioned by our social and economic setting but we are not determined by it. There is an element in man which cannot be accounted for by nature. He can sit in judgement on himself. Pascal refers to this truth when he says that, though the unthinking forces of nature may crush man, they do not know what they are doing but He knows. This is his self-consciousness, his freedom, his superiority to nature. He cannot become a mere function of society. 'The greatest mystery', Mr. Malraux makes one of his characters say in *The Walnut Trees of Altenberg*, 'is not that we have been flung at random between the profusion of the earth and the galaxy of the stars, but that in this prison we can fashion images of ourselves sufficiently powerful to deny our own nothingness.' All these point to the creative role of human choice and responsibility.

Regarding the future there is nothing sure or predestined or guaranteed. The only certainty is that the good will prevail over evil or, as our national motto says, 'Truth will triumph, not falsehood.' The spirit in man is the source of his freedom. We are free to choose the cause. When once we exercise our choice, we may not be able to alter or affect the effect.

History tells us that all those who chose world domination by force broke against the rock of moral law and came to their end. The Indians and the Greeks have a saying: 'Those whom the gods would destroy they first make mad.' The Pharaohs, the Caesars, and the Hitlers had tried force and failed. We are not asked to speak the truth or love our neighbour only on condition that the other person does it. A great nation has to take courage and lead the way.

We are yet in an early hour of the morning of man's history. His civilized life is hardly ten thousand years old and even in that period many empires and civilizations have appeared and disappeared. We see failures and collapses, recoveries and victories. We need not assume that our civilization is the final expression of human wisdom. If we believe in the moral government of the world, we will adhere to the principles of love and brotherhood, and our civilization will survive. It will disappear if it does not conform to them.

We have advanced across the centuries. We believe in creative evolution. The only absolute in the world, Marx says, 'is change'. Let the dead bury the dead. The past of exclusive nationalism, racial superiority is a burden. What counts is the future, the promise. We have scrapped bows and arrows, wheels and carts. We need today the courage to change, to scrap old prejudices, old approaches to international problems. Even as we have adjusted ourselves to the machine age, we have to adjust ourselves to the nuclear age.

* * *

The crisis that is facing us is not a material or an economic crisis; it is not an intellectual crisis. It is a moral and spiritual crisis. We have mastered the forces of nature. We can produce food from the inexhaustible plant life of the sea. Power released from the atom and drawn from the sun and the stars can lift from the backs of men the burdens they have borne for centuries. We have discovered remedies for deadly diseases of body and mind. By harnessing new energies we can raise human well-being to undreamed of levels. We can produce enough material goods to provide for all the people of the world. The old causes of war, hunger, poverty and hopelessness need not be there. As men are freed from the battle of physical existence, they will press forward in their struggle against ignorance, suspicion, malice and hatred. Our intellectual achievements are unparalleled.

If, in spite of these possibilities of material abundance and intellectual power, peace is still in peril, it is due to a cussedness in human nature which still persists, a moral blindness, a spiritual affliction which we are unable to overcome. We have not yet learnt to behave as members of an international community, in spite of our membership of the U.N. We are suffering from a split mind. Not only has the atom been split in our time but our minds, hearts and consciences are rent as under.

In spite of our great advances in science and technology, we are not far removed from the brute. Animals squabble occasionally and fight ferociously at the mating season but they do not periodically destroy each other.

It seems simpler and easier to remake the face of the material environment. To remake our inward nature, to accept the values of spirit which make for creative living, justice, freedom and equality is a harder task. In a letter to Joseph Priestley, discoverer of oxygen, Benjamin Franklin expressed his conviction about the limitless progress to be expected of science and added the following hope: 'O that moral science were in a fair way of improvement, that men would cease to be wolves to one another and that human beings would at length learn what they now improperly call humanity.' The Kingdom of Heaven within us is struggling against the brute forces of the world. The evil in us has to be conquered. What we need is the inner development of man himself, of man as a spiritually, morally and socially creative being. We need a change of heart, a conversion of the soul.

In our generation we have had two world wars. We won the wars but lost the peace. The victors were unable to find the patience necessary for reconciliation. After the First World War we set up the League of Nations but it failed on account of our nationalist obsessions. After the Second World War we set up the United Nations Organisation with the objective of maintaining peace by removing the causes of international tension and creating an international order based on justice, freedom and tolerance. Its work is greatly hampered by its lack of universality and the division into various power camps. It is this division that carries the threat of war which weighs on all people of the world. There is tension between the ideal of international order and the facts of international politics. The prospects of peace are bound up with co-operation among the great powers.

How can we overcome the present division in the world? We cannot adopt the Calvinist view that there is only one truth and those who think differently are inspired by the devil. If we pose the problem in terms of absolute good and evil, if we create pictures in black and white, between religious faith and materialistic atheism, between freedom and tyranny, we make communication and understanding difficult. Each one says that the love of truth constrains him to denounce error.

The human individual, the strange and significant product of evolution, must think of the human family as one. He must look upon his fellowman as a friend and collaborator in a common purpose and not as a potential enemy to be feared, hated and killed. He should not assume that he has eternal and ultimate wisdom and the other absolute folly. No portion of mankind was as good as it thought itself or as bad as it was thought by its enemies. We must tread another road in dealing with human relationships.

'Public opinion in America has changed in regard to the nuclear weapons. When only this country had it, it was accepted as good since it was capable of stopping a great war and saving millions of lives. When other countries also developed it, we found its real character as a dangerous weapon which might destroy civilization. What was once regarded as acceptable is now treated as evil.

When we speak of a 'free' world, we are using the word 'free' in a loose sense. It includes a number of unfree military dictatorships, non-democratic authoritarian government, several of which exalt race discrimination. Hitler is said to have proclaimed once: 'I am making all Germans unfree in order to make Germany free.'

* * *

Modern psychiatry tells us that it is no use becoming furious at people who behave stupidly or wickedly. Instead of losing our tempers with them, we should study the reasons for their behaviour. Indignation against wrong is better than indifference to wrong. But gentleness and compassion are better than both. There is a temptation to look upon our opponents as inhuman monsters, infernal fiends plotting in dark cellars, who require to be destroyed for the safety and health of the world. But we should try to change the mind of our opponents without resorting to force. If we disagree with our opponent, it is not a justification for destroying him. If he is blind, we should help him to see. We must awaken the sense of justice in him. We should never weary in our effort to help him to cast off his error. Then what is good in him will unite with what is good in us and we will all march unitedly towards the goal. Cruel methods are not necessary even to drive out cruelty. We should not use the methods of the devil to drive out the devil. Not condemnation of each other but mutual education is necessary. Vengeance is God's.

The present situation calls for repentance, a recognition of our imperfection and fallibility. It is not a question of doves in some countries or vultures in others or vice versa. The spirit of God transcends man-made curtains. The first step we have to take is to look upon our enemies as people like ourselves who are anxious to lead quiet, respectable lives. They are like ourselves, fathers, mothers, sons and daughters and children. They are men, like ourselves with energy and sacrifice, eager to rectify the injustices of the world, real or imaginary. We can then understand what the reasons are for their behaviour which is so disagreeable to us. We shall then realize that their success is the measure of our failure.

If we are blameless we can judge others, but we are not. At the Mormon Tabernacle at Salt Lake City, I heard a Spiritual which opened with the line, 'Were you there when they crucified the Lord?' We are here today when we are crucifying Him on the cross of power, domination, national idolatry and racial arrogance. We see man's capacity for monstrous deeds, the susceptibility of even the gifted to delusion and of the seemingly decent persons for malice and hatred. Out of the heart of our civilization came the two World Wars, persecution of millions of people on racial grounds, concentration camps, torture chambers and atomic destruction. We acquiesced in all these and even now do not seem prepared to put our ideals and insights into practice. We seem to be satisfied with dispensing advice and indulging in self-righteous warnings. If we wish to stop subversive movements the flame of social reform should burn in our hearts. We must bring the light of a new life into the minds and hearts of millions of underprivileged men, women and children.

If we want to convert our opponent, we must not always speak of his lower side, his defects and shortcomings. We must present to him his own higher and nobler side. Systems which involve millions of people cannot be explained by simple formulas of political machinations and intrigues. We must discern a human mind and a human heart even in the fanatic faces of our opponents who adopt different ideologies.

Professor Arthur H. Compton wrote when Soviet Union was still in existence:

> The achievements of Russian production in turning out military weapons has won our respect and admiration. Their planes are excellent, in some respect superior to our own. Their guns are of the very highest quality. The speed of their development of atomic weapons has taken us by surprise. Their advance toward practical use of atomic power means that they will compete strongly for world leadership in the field.
>
> In education a similar growth has occurred. A generation ago the Russian people were largely illiterate. Today they are graduating nearly twice as many engineers as is the United States. In music and drama they have a notable record.
>
> Such educational development cannot occur without giving a better understanding of the value of freedom. In this regard we can be confident that time is working for us. Already we have seen substantial changes occurring. These changes have not been altogether political. The fall of T.D. Lysenko from his position as dictator of a false science of genetics is an example. The fact is that Russia is becoming industrially and intellectually a modern nation. Whether by gradual change or by violent internal upset, we can rely on the universal social forces to bring that great half of the world into such a condition that harmonious adjustment with the free half becomes possible. And social changes are occurring in Russia at a very fast tempo.[2]

[2] *Atomic Quesst* (1956), pp. 351-352.

Science and technology have relieved mankind of degrading drudgery. They have added to man's comfort, health and enjoyment of existence. The average life span of man on earth has been greatly increased. Science has added to the dignity and stature of the individual. When man is relieved from the battle for physical safety, he becomes a little creator. Every challenge of science has added to man's moral stature. As we find that the world is much more wonderful than we ever dreamt it to be, we are led into new fields of awareness, new ranges of attainment, new realization of destiny. New knowledge is both a challenge and an opportunity. 'For it is not yet known what man may be.' He must, without haste and without rest, strive to reach the quality of human greatness, that is, greatness in humanity. Mankind is the higher sense of the planet. When it realizes its destiny of inward awareness and social compassion, then the great reconciliation among the peoples of the world in which all groups win and no one loses will take place.

If we wish to secure enduring peace, pacts and treaties are not enough. The two World Wars started with the repudiation of pacts and pledges. Peace is not the result of an armed truce or a cold war. It is possible only if all nations spontaneously accept the ideals of justice, freedom and decency, justice for all nations, freedom for all men of goodwill and decency in our international behaviour. The new civilization is open to all people who believe in fundamental values, in spite of race or creed. The limits of the community are world-wide and are decided by attitudes of mind and not frontiers of geography. Our civilization is not altogether bankrupt. It can yet lead us to a freer, juster, friendlier world.

If the disordered pattern of our society is to be set right, we have to fashion a peace that will assure justice and individual liberty to all and remove the injustices which are the sources of unrest and conflict. To establish peace, we must remove the conditions which make for wars.

We are living in a period of great change and rising aspiration. There are some forces in human nature which cannot be destroyed, such as love of one's soil. Nationalism is a decisive factor in modern history. It can be kept pure only if the nationals of a country have a human feeling for all men. From the love of our country we must progress to a love of mankind. It should not develop into a hatred of other nations. No nation should desire to promote its interests at the expense of others. Genuine nationalism is consistent with membership in an international order. The principle of variety in unity should guide the behaviour of all nations. If we suppress national aspirations and support feudalism and reaction, we lose the battle. We must assist subject nations to win their freedom. The U.N. Organization should devise machinery for effecting peaceful changes and completing the unfinished movements for independence.

It is our duty to promote racial harmony. Racial injustice is the raw material of strife. Human beings should not be deprived of their dignity on account of their race or colour. Men of social conscience are everywhere striving to remove racial discrimination. If we acquiesce in racial segregation, we participate in something unclean. Enlightened self-interest demands the removal of this injustice as soon as possible, It is not merely self-interest. It is fellow-feeling. In our country we have troubles about caste discrimination and communalism which we are striving to eliminate.

Again, internationalism should be not only political but economic also. Even as within a nation the more fortunate persons are called upon to assist the less fortunate ones, so in the international world the more favoured ones should assist the less advanced. We have now food surpluses in some countries and starving peoples in others. All men have a right to be fed, clothed and sheltered. Their minds should be trained and their spirits nourished. Backward nations should be helped to acquire the economic apparatus by which they can raise themselves. Our fight should be against hunger, disease and illiteracy. It is possible to free humanity from these scourges. If we do not, the revolution of the destitute and the desperate will shake the world.

Those who suffer from privation and poverty are attracted by other alternatives which can hardly worsen their position. A sense of hopelessness among the people is a potent cause of upheavals. People will put up with any amount of political terror if they are earning and eating. Millions in Asia and Africa are intent on improving their material condition; until that happens they will not find the values of freedom. Human beings generally resort to violence when pressed by economic want. The unequal distribution of power and wealth, the wide differences of health and education among the nations of mankind, are the sources of discord in the modern world, its greatest challenge — and if unchecked, its greatest danger. We should look upon the world as one unit. We need a world economic development programme. If we do not wish to destroy the world, advanced nations should set apart a small fraction of their national income for the purpose of this programme. We have the insight. Why do we not have the zeal? When we know that the future of underprivileged countries is unpredictable, why is it that we do not have a sense of urgency, conviction, passion, warmth? Why are we not stirred by the idea of one world which will compel us to liberate the poor and the exploited strata of our society?

If we assist subject nations to become free, if racial determination is removed and if underprivileged countries are helped to raise their standards of living, the root causes of conflict will disappear.

* * *

An international society is the goal to which we are moving. We cannot reverse the processes of history. There must, however, be a machinery for enforcing the rule of law among the nations of the world. In a disarmed world we should have an effective United Nations with a police force universally recognized and respected. While it should not interfere with the internal affairs of nations, it should protect all nations against lawlessness and aggression from outside. No single nation has the right to police the world but all nations can contribute their equitable share to the U.N. force. We should transcend in some measure our national sovereignty for the sake of the survival of the human race even as individuals have given up the right to enforce their will by force.

Common interests transcend differences of ideology. We can mobilize the consciences of our contemporaries, even those living in communist countries, for shaping a better future for humanity. The peoples and governments of the world, whatever their ideologies may be, are interested, in this age of the conquest of outer space, in preserving peace and averting war. There is no magic formula or shortcut solution to peace. We may proceed step by step, reach limited agreements, improve the political climate, strengthen the confidence among nations and foster policies of live and let live, of co-existence. It is the only way to develop a moral community in which we can live together and work for a fuller life of our peoples and remove the greatest fear, which is fear itself.

It is therefore essential for the leaders of the great Powers to start negotiations in an atmosphere of goodwill, sincerity and imagination, with a determination to reach agreement. Whatever the difficulties may be, the search for reaching agreement should not be abandoned.

Many years ago, Alexander Hamilton wrote to Mr. Rufus King who was American Minister in Britain: 'This country will ere long assume an attitude correspondent with its great destinies — majestic, efficient and operative of great things. A noble career lies before it.' Leadership cannot be exercised by the weak: *nāvam ātmā balahīnena labhyaḥ*.

The way of peace requires that men and nations should recognize their common humanity and use weapons of integrity, reason, patience, understanding and love. There are many forces at work which give us hope and assurance. Even if we fail, we should not give up our efforts. Failure we have with us always, but man possesses an unconquerable self that through failure and tragedy may rise to higher reaches of spiritual victory by the transcendence of evil.

Man must persist in his belief that the incomprehensible is comprehensible, otherwise he would cease to explore.

CHAPTER 5

THE METAPHYSICAL QUEST

The metaphysical temperament, which seeks to penetrate behind the limited knowledge which comes to us through sense perception and interpret the nature of the world by means of general ideas has characterized the Indian mind from the beginning of its history. The German mind has been reputed for its profundity of thought, imaginative power and the capacity to probe into the depths of human experience. There are two paradoxical aphorisms of Goethe which reveal the inquiring spirit of the German mind. 'Man must persist in his belief that the incomprehensible is comprehensible, otherwise he would cease to explore.' 'The highest happiness of a rational being is to have explored what is explorable and quietly to revere what is unexplorable.' To attempt to see things, persons and events in their inter-relationship is inevitable to the human mind. Metaphysical ideas are founded on a basic awareness of what is implied in experience and cannot be altogether justified by scientific measurement of rational logic. We cannot attain clarity in regard to them. Those who were greatly impressed by the triumphs of science found systems of metaphysics irrelevant, if not meaningless. While the scientifically inclined were hostile to metaphysical views, the religiously inclined felt that the prevalent systems of speculative idealism dwarfed the nature of the human individual and permitted the intrusion of reason into the realm of faith. The individual loses his significance if the whole universe is the march of the Absolute. So protests were made against the doctrine of Absolute Idealism in the interests of scientific knowledge and the significance of the individual. Science and faith were both opposed to metaphysics.

Auguste Comte (1798-1857) and Sorën Kierkegaard (1813-1855) appeared about the same time. But their hostile reactions to Hegelian Idealism were not taken seriously at the time. They have now become articulate and attractive to many thinking people. The two wars which we had in one generation, the political and social upheavals through which we passed, the general misery and hopelessness of life have resulted in a state of disruption, disillusionment and despair. In the state of intellectual confusion and spiritual prostration, we are taking to all sorts of remedies, theological orthodoxy (Karl Barth), neo-scholasticism (Jacques Maritain and Etienne Gilson), Marxist Materialism, Logical Positivism or Nihilistic Existentialism. We wish to be saved, at any cost, from a sense of lostness, homelessness and these systems profess to give us a sense of belonging.

Sooner or later we have to come to grips with life, find our bearings and reach certain beliefs and values. It is not a question of whether we should have a metaphysics or not for we all have one. The question is whether it is to be an unexamined and even unconscious metaphysics, or a system of ordered thought which is deliberately achieved. Even systems of Logical Positivism and Existentialism, which revolt against metaphysics, set forth metaphysical views.

* * *

Comte claimed that there are three stages of mental development. In the first stage the theological men seek to explain the world in terms of myths. In the second, the metaphysical men seek to account for things by devising speculative, rationalistic theories. But the third stage, the positivist, is that of modern science, in which we are content to trace, discover and record what is observed to happen without trying to go beyond facts. Metaphysics for Comte was pre-scientific and is superseded by science. The positivism developed by Comte was not a theory but a method, which sought in every field for scientific explanations instead of theological dogmas or metaphysical speculations.

Logical Positivism wishes to adopt the empirical method. Its revolt against idealism, did not in the early stages, mean a revolt against metaphysics. G.E. Moore, for example, said in 1910 that 'the most important and interesting thing which philosophers have tried to do was to give a general description of the whole of the universe.'[1] Bertrand Russell announced in 1918 his intention of setting forth 'a certain kind of metaphysics.'[2] He developed a system of metaphysical realism. Paul Natorp (1854-1924), the co-founder of the Marburg School,

[1] *Some Main Problems of Philosophy* (1953), p. 1.
[2] *The Philosophy of Logical Atomism.*

speaks of an ascent from the world of contradiction to a region of pure harmony and unrestricted affirmation.

Ludwig Wittgenstein became the starting point of two different schools of philosophy. Linguistic Analysis and Logical Positivism. He tells us that all true propositions are contained in the natural sciences and philosophy cannot make any true assertions on its own account. Its sole function is to clarify our thoughts.

The function of philosophy is the analysis of science and its logic and language. We cannot seriously consider questions of philosophy unless they are precisely stated. But Linguistic Analysis is itself an approach to metaphysical thinking. Logical Positivism has great respect for science and mathematics and has developed a distaste for metaphysics. It affirms that metaphysics is in principle impossible or meaningless. It adopts the verification principle as the criterion of meaningfulness or significance. The meaning of any statement is evident from the way in which it could be verified. It is assumed that verification must always terminate in empirical observation or sense experience. The only exceptions are the analytic formulas as those of mathematics which do not require to be empirically verified. The statements of metaphysicians and theologians, and moral and aesthetic judgments could not be empirically verified and therefore they are non-significant, meaningless.

Only statements require to be empirically verified are not commands, entreaties, promises, interjections or expressions of intention.

Many difficulties arise here. First, positivists themselves have expressed metaphysical convictions like the doctrine of physicalism. When we speak of rivers and mountains, flowers and fruits, birds and animals, when we classify the objects of the world, we are on the metaphysical track leaning to empirical realism. When science speaks of the world as having a continuous structure and uses the concepts of nature, cause, law, it is making metaphysical assumptions about the structure and continuity of nature. What gives value to scientific discoveries is derived from a source other than science itself. Logical Positivism which rejects the validity of metaphysics on the ground that metaphysical statements are incapable of empirical verification, is itself a metaphysical view that sense experience is all, and any statement that cannot be tested by it is unintelligible, meaningless.

Is experience limited to sense experience? We cannot deny the experience of purpose, of choice, of vision, of beauty, of apprehension, of truth, though they may not be capable of scientific measurement.

Husserl's phenomenology is an attempt to be as empirical and realistic about the world of thought as other empiricists have been about the world of sense-perception. Husserl holds that the ideas of which we are aware in thought are objectively real even as are the things with which we are in contact in sense-perception. If the latter *exist* as we perceive them, the former *subsist* as we think them. We must develop that intensity of intellectual vision by which we shall see the ideas which subsist in the world of thought as they really are, in all the detail of their logical structure.

Again we have experience of values, of the world of spirit. We speak of *darśana*, seeing, vision, when the soul is penetrated by a Being that is immensely more powerful than itself. The soul turns inwards and concentrates in the central part of its being when withdrawn from body and space, beyond relation and time it enters into the presence of Divine Creativity. This is not a rare or privileged event. No one is so poor as not to have felt its light and liberation. Whether we speak of union or communion with

God, or commitment to God, we have an experience which is spiritual, not perceptual or conceptual. It is true that we have no other source of knowledge than experience but this experience is of different kinds, of scientific laws, of moral obligation, of spiritual reality.

Again we have both direct and indirect verification. This building is tall. It can be verified directly; the theory of relativity is verified by its consequences in so far as they are calculated and observed. This is indirect verification. Any metaphysical theory can be verified indirectly by its adequacy to account for the observed facts. Speculative metaphysics is a tenable inquiry with a field of its own.

Indian philosophic thought finds an empirical basis for its conclusions. By an examination of given facts, it constructs a theory of Reality, which has logical consistency and empirical adequacy. Reality is what is ideally intelligible. Bradley says, what *may be*, if it *must be*, it *is*.

Modern philosophic thought is reverting to the ancient concept of the world as *saṁsāra*, perpetual movement, a procession of events which supersede one another. The world we know, which science studies, is a process. It is not complete in its present situation. It derives from the past and moves into the future. Does it reveal another beyond itself? If it does, what is its relation to the process itself? Is this process purposeful? These are questions of metaphysics which we cannot avoid raising, though we may not be able to answer them satisfactorily.

Science tells us that there is regularity, a law discernible in the cosmic process. There is not only movement but the movement seems to have a direction. No situation is exactly identical with the past out of which it has arisen. The element of novelty is not altogether predictable from the past though it may be related to the past which we notice when once it occurs. The cosmic process is not a never-ending repetition of a limited number of situations. The present adds significance to the character of the past and may pave the way for the realization of unexpected possibilities of the future.

If a tree grows from the seed and the child grows from an embryo, we seem to be justified in assuming a purposeful quality in the life experience of these units. Does not the story of the life-process indicate a directional significance? The vast astronomical and geological processes have provided the environment required for the appearance and development of life. From life arises animal mind, from animal mind human intelligence and from it spirit, wisdom, compassion, joy.

The very effort of man to know the secrets of nature, to sit in judgment on it, indicates his participation in the creative process. The scientist's self-dedication and moral concern shows the primacy of the living spirit in him.

Marxism which is aware of the cosmic process and its trends tells us that matter itself is dialectical. But matter and dialectic do not mix. We cannot attribute to matter the dialectical behaviour which Hegel gives to mind. Hegel in his *Dialectic of Nature* observes: 'Our mastery of it (nature) consists in the fact that we have the advantage over all other beings of being able to know and correctly apply its laws.' The human subject is superior to the object. Nature and human existence are not self-explanatory.

The world can be interpreted in terms of the activity of a Being who transcends it. Even as there is continuity between the seed and the tree, the child and the man, there seems to be a continuing identity which endures through all changes of process. This identity finds expression in the variations and changes of the process, in the potentiality of energy, in the direction of biological growth, in social change. The human being enduring through the years is an image of the larger identity of the whole to which he is related. That is why he is capable of participating in the work of that larger Spirit who is the meaning of the universe, discovered at the end though implied from the beginning. The evidence for this ultimate identity is to be discovered not in the interests of knowledge but in the whole process, with its ordered and progressive character. Śaṁkara, for example, tells us that there is a fundamental principle standing behind and beyond the world process which inspires and informs the process which is steadily manifesting new values.

ékasyāpi kūṭasthasya citta-tāratamyāt
jñanaiśvaryāṇām abhivyaktiḥ pareṇa
pareṇa bhūyasī bhavati.

In 1784 Herder published the first part of his *Ideas for a Philosophical History of Mankind.* In it he argues that all things are actuated by a single spiritual principle, which works differently at different levels but is essentially one in its operations. Even Thomas Hardy in his *The Dynasts* saw the world not as a machine but as a living organism moving to a purpose, namely, redemption. The Immanent Will is grouping towards self-knowledge. It has resulted in the evolution of the human soul whose active and increasing self-knowledge is the highest expression of the Universal Force.

Bergson, Alexander, Llyod Morgan and Whitehead developed a view of the universe in which spiritual values have a determining role. Science reveals to us that there is a spiritual presence greater than man. The end of man is to place himself in harmony with this presence.

In India religion and metaphysics were not divorced from each other. In all the Scriptures rational scrutiny of religious thought is insisted on. There can be no retreat from modernity and science. Metaphysical efforts give to religious thought dignity and strength.

* * *

The Existentialist rejection of metaphysics is based on the recognition of the inadequacy and relativity of scientific knowledge. Existentialism is not a specific doctrine. It is a way of thinking which takes self-conscious existence as the proper subject and point of departure for philosophy. To exist is to be a self-conscious being vividly aware of himself and engaged in a great personal adventure.

Existentialism attacks the idealist position which looks upon world history as the march of the Absolute Spirit in which the human individual has little part. But all historical acts are acts of men. Men make history.

Existentialism is empirical in its method. While the older empiricists applied their belief to nature, Existentialists apply it to human existence. The *Upaniṣads* ask us to know the self: *ātmānam viddhi.* The *Bhagavadgītā* says that of all types of knowledge the knowledge of the Self is the most important. We must achieve a profound understanding of what is meant to be human. The knowledge aimed at is not psychological knowledge which is after all a species of scientific knowledge. Psychology gives us the mechanism and conditions of existence. It breaks man into a series of fragments which are studied by the different sciences. Man is treated as an object among objects, emptied of spiritual, orientation and moral certitude. Instead of individuals, science gives us concepts, doctrines. Man becomes a collection of ideas, feelings and desires. A tennis ball is an object but man is both subject and object to himself. Existentialism attempts to study the meaning and values of existence. For this we have to pass beyond science.

Religions, especially Hinduism and Buddhism, describe the broad features of human life, the pervasiveness of suffering in a way not wholly alien to the doctrine of the Existentialists. The world is subject to time, historicity, change. Life is haunted by death, beauty by decay. Nothing abides; everything passes away. What remedy avails against this malady of mankind? The *Upaniṣad* writer prays: 'Lead me from the unreal to the real; lead me from darkness to light; lead me from death to immortality. Lead me from the world of time to the reality of eternal life.' The Buddha saw the world to be full of suffering, sickness, old age and death. The last of the four sights he encountered was that of a recluse, with a radiant and tranquil face, who answered the Buddha's question by stating that he was in search of liberation from time, being afraid of birth and death. Indian thinkers take birth and death and not merely death as symbolic of time. Impressed by the vanity of our projects, the futility of our achievements, the restlessness of temporal life, its confusions and contradictions, its ultimate nothingness, the Buddha tells us that each one has to pass through it all in order to fulfil himself and recognize at the depth of all struggle the lasting peace of nirvāna. The Buddhist asks: Why am I bound to be what I am? Why should I wear the mask of this personality and endure its destiny with all its limitations and delusions? How can I attain another state, that of not being anything particular beset by limitations and qualities that restrict my pure unbounded being? St. Paul asks: Who shall deliver me from the body of death? He announces that 'the last enemy to be destroyed is death.' Augustine speaks to us of the ceaseless unrest which marks the temporal life of the individual.

The psychological dissatisfaction with the meaninglessness of the world is the stimulus to the metaphysical quest. Man is dissatisfied with his finitude because he has infinite longings. His ontological interest is a challenge to transcend the human predicament and seek another world, a new level of being. Man becomes aware of his finitude because of his potential infinity. When he is aware of his existential situation, he becomes aware of the power of Being in him. When he is aware of his sin, this consciousness is the shadow of the absolute standard in him. It is a demand which the soul makes on itself. It is not God's judgment but man's own judgment on himself. He knows and applies the absolute standard because he participates in the nature of the Absolute. According to *Advaita Vedānta*, the reality of Brahman the Supreme Being does not need to be proved since it is a datum of consciousness, bound up with the consciousness of man's own existence.

Being is the answer to meaninglessness. Being does not have its consummation in nothingness. Death is not all. The transition from becoming to Being is an attempt at the restoration of metaphysics. Religious life is the life lived in the power of Being. It fosters and develops the metaphysical urge and ministers to the need in man for adoration.

In recent European thought, Kant may be treated as the forerunner of both Positivism and Existentialism. At the beginning of the *Critique of Pure Reason* he says that because our knowledge arises in experience, that is no proof that it is derived from experience. Kant distinguishes three levels of cognitive activity; the Aesthetic with the forms of perception, the Analytic with the categories of the understanding, and the Dialectic with the Ideas of Reason. The categories of the understanding are *a priori* conceptions, structural tendencies of the mind without which we cannot have knowledge of sensible phenomena. They are not logical abstractions but active manifestations of the unifying principle of mind. They apply to objects of sense as conforming to the universal conditions of a possible experience, phenomena, and not to things as such, noumena. A transcendent use of these *a priori* principles is illegitimate.

While the categories of the understanding are immanent, that is, adequately realized in sense-experience, the Ideas of Reason are transcendent. No objects can be presented in experience that are adequate to them, the subject, the supersensible substance from which conscious phenomena derive, the object, the world, the totality of external phenomena and God, the union of subject and object, the source and unity of all existence. Though the Ideas of Reason have no objects in a possible experience, they are yet the ideals of all experience. We cannot apply the categories of the understanding to them. The soul, the world and God are not substances and causes. If we apply the categories, we get the pseudo-sciences of Rational Psychology, Rational Cosmology and Rational Theology. The Ideas indicate the aspirations of thought, the demands and dreams which we cannot relinquish. There is no science of objects answering to the Ideas of Reason, though we are obliged to act as if there were such objects. Our cognitive activity rests on a faith and a hope.

Everything empirical is conditioned and relative while the Ideas are absolute and unconditioned. Moral life gives a deeper meaning to the Ideas of Reason. The intuitive apprehension of the moral law is different from the logical apprehension of objects in space and time. Kant tells us that we can have apprehension of the unconditioned.

After we have denied the power of speculative reason to make any progress in the sphere of the super-sensible it still remains to be considered whether data do not exist in our practical cognition which enables us to render determinate the transcendent character of the unconditioned, and in this way, as metaphysics seeks, to over-step the bounds of all possible experience with our *a priori* knowledge — which knowledge, however, is valid only from the practical point of view.

What metaphysics tries in vain to prove by purely intellectual means is all the time accepted by the ordinary man on the basis of practical or moral experience. It is not an object of scientific knowledge but of rational faith. In this inaugural Dissertation, Kant speaks of *intuitus intellectualis*. The ideals of truth, goodness and beauty are the expressions of the Spirit in us. Their objects are ontological, the very substance of being. The contents of spiritual consciousness are the conditions of human knowledge, morality and aesthetic life. The unconditioned principles are more Ideas of Spirit than of Reason. Sense, Understanding and Reason are the three ways in which the Spirit in us functions. The appeal of metaphysics is to a judgment more basic than either sense-experience or national logic. It attempts to assess the reasons for and the limitations implicit in the presuppositions of science and logic.

Kierkegaard speaks of the existential dread but the dread is what spurs the individual to that leap of despair which is faith. It is the Dread of the soul suspended between finite existence and its infinite possibility. Kierkegaard's work *The Concept of Dread* ends with the words: 'So soon as psychology has finished with dread it has nothing to do but to deliver it over to dogmatics.[3] When Kierkegaard speaks of the power of *angst* over existence, it is with the intention of helping man to escape from that power. Man's awareness of *angst* makes the world for him a desolation. His analysis of human existence was made with an apologetic purpose. Kierkegaard affirms that the transition from psychology to

[3] E.T. by Walter Lowerie (1946), p. 145.

dogma is always an act of faith and not one of logical necessity. He denounces Metaphysical system-building as found in Hegel, in the same way in which Luther protests against the intrusion of Aristotelian philosophy into the Kingdom of Faith.[4] For Kierkegaard man's relationship to God is what makes him human.

The great German thinker Paul Natorp reminds us of an Indian teacher who adopts silence as the best expression of the great mystery. Truth exists by its own majesty. Its language is silence. When we sit near a seer, he does not deliver a message but we sense the consuming heat and the kindling light of his spirit. He creates a mood, a temper rather than a conviction or a belief. To be born again and become as a little child was for Natorp the highest goal a man can reach and he feels that nobody has achieved it, not even Jesus. This religious background determines the philosophical ascent from the world of confusion and contradiction to a region of pure harmony and unrestricted affirmation.[5]

---

[4] Martin Luther writer: 'All the articles of our Christian belief are, when considered rationally, just as impossible and mendacious and preposterous. Faith, however, is completely abreast of the situation. It grips reason by the throat, and strangles the beast.' On this Karl Barth makes the remark: 'He who can hear this, let him hear it, for it is the beginning and end of history.'—Professor H.J. paton: *The Modern Predicament* (1955), pp. 119-120.

[5] Bradley looked upon the universe as fragmentary, disjoined, irrational. He pointed, out that everything which is taken to be true or real is self-contradictory and therefore, dismissed as mere appearance. It is quite alien to the nature of Reality whitch is to be found only in the Absolute, which neither is nor contains a number of things. In the Absolute all separation is overcome, all distinctions vanish, all relations are merged.

Sartre's Existentialism, though avowedly atheistic, lends considerable support to a spiritual view of the universe. He gives us a way of thought which seeks to be baptized into faith. He affirms that even if God did exist, man would still have to be his own saviour. We should be brave enough to face reality and not accept consoling myths contrived to give us peace of mind. We must be prepared to live with *angst.*

Albert Camus tells us that the myth of Sisyphus[6] teaches us the hopelessness and futility of man's situation. We have to roll the rock of existence without any hope of rest or of finding any meaning in our life. He calls man's relation to the universe as Absurd but he contends that this relation is to be fulfilled through courage and reason on man's part. This is, however, a sign of man's faith in himself.

For Martin Heidegger, authentic existence is one in which man finds himself and unauthentic existence is one in which the individual is lost and scatters himself in the world. He is uprooted when he falls into the world; he is not then disturbed by the ultimate issues of existence. When his security is taken away from him the mood of anxiety breaks in to shatter his contentment. He has then a sense of alienation, that he is cut off from his true self and its authentic possibilities. Evil is a falling away of man from himself, from his authentic being. When the original possibilities are lost or are rendered inaccessible, we have a state of fallenness. The deeper his fall into the world the further is he from himself. To restore authenticity means to unify the scattered self, so that it is withdrawn from false concerns and stands in its original possibilities. Conscience is the call of the authentic self to the fallen self. For Heidegger we approach most closely and intimately to reality in human existence. By analysing self-consciousness where reality discloses itself to us, we

---

[6] *The Myth of Sisyphus and Other Essays.* E.T. by Justin O' Brien (1955).

will attain knowledge of the objective nature of things. By profound metaphysical thinking we can overcome the self-estrangement of our existence, get reconciled to reality and give meaning and hope to our existence.

For Karl Jaspers, existence is not an idea but the most concrete form of experience. It has not the character which abstract ideas have. To know oneself as a finite being in a world which conditions and restricts one's liberty is to know oneself as transcended. All that exists tries to get beyond itself. 'Man cannot remain himself except by living in a relationship with the Transcendent.' When Jaspers insist that man is 'open to the Transcendent,' he is referring to the capacity of man to transcend himself. The conviction of the meaninglessness of life leads us to a new being which promises creativity, meaning and hope. There is need for believing that below the actuality of existence there is the mystery of Being.

Philosophy is said to be a *darśana*, a vision, a new way of seeing beyond the horizons of time. When this seeing, which is a sudden growth in understanding, occurs our whole outlook on the world is transformed. When Bergson defines metaphysics as 'the science which claims to dispense with symbols' he means that it seeks the intuitive vision of the Real beyond the distorting forms of conceptional thought.

Religion is not a philosophical proposition, not a historical life. It is a personal discovery that the apparently indifferent world conceals as its reality an intimate concern for each individual, as of parent for child.[7] 'Behold, I stand at the door and knock; if any man hear my voice, and open the door, I will come in to him and will sup with him and he with me.'

Eternal life does not take us away from the world of time. It reveals to us the world as a reflection of the Divine

[7] *Bhgavadgītā*, XI, 44.

splendour and the bond of its unity. The world is not a threat to man's authentic being. We have to live in the world with the perspective of the eternal. Man's true life is not one of material security. His life then is one built on sand. His authentic existence of understanding and love is one built on rock. Man is essentially a creator dealing with existence which aspires to Being. He is not a lonely ego in an impersonal natural order. He becomes a pulsating centre of action by the submission of his will to the Transcendent. By his submission the soul becomes both acting and acted. Its freedom coincides with the Divine Activity. Its sole end is to prefect the creation of the human species. When charged with the vision of unity, the self returns to the world of action, bringing with it a new comprehension of its relation to others, seeing both itself and them as personal entities bound together in a process of mutual creation. Its activity has a religious significance and must be understood in the light of a universal and divinely inspired creative process, whose end is the union of all human beings.

This view is the only answer to the casual character of our lives where we do not have any significant purpose, where we are alarmed by life's risks and uncertainties, where we do not have faith in the future.

The revolt of modern philosophy in its Positivist and Existentialist forms has been a healthy and liberating influence. But we cannot rest content with revolt. We need constructive philosophy, an articulation of ultimate presuppositions about the world we live in. This is possible only by hard metaphysical thinking. Faith has to be a rational one. Metaphysics is not the enemy of faith. It alone can restore to men the spiritual wholeness which they seek to attain but fail to do. It is the metaphysical effort that gives dignity to the human species. No culture can last unless it supports this effort and encourages the confidence that man is capable of insight into the nature of the process in which he participates.

The self is deeper than the perceptions, thoughts and feelings.

CHAPTER 6

THE ASIAN VIEW

The Asian view of man is not very much different from the ancient European view of man. I do not believe in the pseudo-science of national or continental psychology which affirms that all Asians are this and all Europeans are that. The history of any people is slightly more complicated than these sweeping statements would suggest. As a matter of fact, the Asian and the European peoples had common beginnings and developed from them relatively independent views and acquired certain features which marked them from each other.

In spite of varying developments, the different peoples of Asia possess a number of features in common, which will justify our speaking of an Asian view of man. This view is essentially a religious one. All the living faiths of mankind had their origin in Asia: Confucianism and Taoism in China; Hinduism, Buddhism, Jainism, Sikhism in India; Zoroastrianism in Iran; Judaism and Christianity in Palestine; and Islam in Arabia. The religions adopted by the Western people are all derived from Asia. In a short space it will not be possible to deal in detail with the different religious developments. I shall content myself with a statement of the Indian point of view, with which I happen to be somewhat familiar. Besides, Indian culture has influenced a large part of Asia's thought and art and affected other parts of the world also. Peoples of different races, languages and cultures met on the soil of India; and, though we read of occasional clashes, they have settled down as members of a common civilization whose primary characteristics are faith in an unseen reality, of which all life is a manifestation, the primacy of spiritual experience, a rigid adherence to intellectual norms, and an anxiety for harmonizing apparent opposites.

The one doctrine by which Indian culture is best known to the outside world is that of *tat tvam asi.* The eternal is in one's self. The Real which is the inmost of all things is the essence of one's own soul. The sage whose passions are at rest sees within himself the majesty of the great Real. Because there is the reflection of the Divine in man, the individual becomes sacred. If we try to possess man as flesh or as mind to be moulded, we fail to recognize that he is essentially the unseizable who bears the image and likeness of God and is not the product of natural necessity. Man is not something thrown off, as it were in a cosmic whirl. As a spiritual being, he is lifted above the level of the natural and the social world. He is essentially subject, not object. Modern Existentialism points out that a type of thought which dominates the treatment of objects is inadequate to the thinker, the existing individual. His inward reality is not to be equated with the qualities by which he is defined or the external relations by which he is bound. We know the self not in the sense we know the object. When we look inwards we find a limit to our knowledge of the inner life. The self is deeper than the perceptions, thoughts and feelings. We cannot see it or define it, for it is that which does the seeing and the defining. It is the eye which is not the object but the subject of our knowing. It can be grasped, not by thought, but by our whole being. Then we realize the existential presence of the ultimate reality in each individual.

The Indian classic, the *Bhagavadgītā*, speaks of the spirit of man as immortal. Weapons do not cleave the Self, fire does not burn him, waters do not make him wet, nor does the wind make him dry. He is uncleavable, he cannot be burnt, he can be neither wetted not dried; he is eternal, all-pervading, unchanging, immovable; he is the same forever.

The term 'personality' is derived from the latin word *persona*, which means literally the mask that is worn over the face by the actor on the stage, the mask through which he sounds his part. The actor is an unknown, anonymous being who remains intrinsically aloof from the play. He is unconcerned with the enacted sufferings and passions. The real being is concealed, shrouded, veiled in the costumes of the play. To break from the confines of personality into the unfathomed reaches of his true being requires disciplined effort. By penetrating through the layers of the manifest personality, the individual arrives at the unconcerned actor of life. Man is more than the sum of his appearances. When Crito asks: 'In what way shall we bury you, Socrates?' Socrates answers: 'In any way you like, but first catch me, the real me. Be of good cheer, my dear Crito, and say that you are burying my body only, and do with that whatever is usual and what you think best.'

The Indian thinkers do not oppose nature to spirit. When the natural life of man comes to itself, his spiritual being becomes manifest. Man's final growth rests with himself. His future is not solely determined, like that of other animals, by his biological past. It is controlled by his own plans for the universe. Man is not an insignificant speck in a depersonalized universe. When we overlook the inward subjectivity of man, lose ourselves in the world, we confuse being with having; we flounder in possessions as in a dark, suffocating bog, wasting our energies, not on life, but on things. Instead of using our houses, our wealth, and our other possession, we let them possess and use us; we thus become lost to the life of spirit and are soulless. It is attachment to nature that is inconsistent with spiritual dignity. It is not necessary for us to throw off the limitations of nature. Our bodies are the temples of the Divine. They are the means for the realization of value, *dharma-sādhana*. When human beings are most clearly aware, most awake, they feel that in some sense which cannot be clearly articulated, they are the instruments for the expression of the spirit, vessels of the spirit. When we realize this, we outgrow individualism, we see that we and our fellow-men are the expressions of the same spirit; the distinctions of race and colour, religion and nation become relative contingencies. We are reminded of Socrates' death-bed statement: 'I am not an Athenian or a Greek but a citizen of the world.' To the large-hearted, all men are brothers in blood, says a well-known Sanskrit verse. The *Bhagavadgītā* tells us that a truly religious man sees with equality everything in the image of his own self, whether in pleasure or in pain.

From the emphasis on the immanence of the Divine in man, it follows that there is not one single individual, however criminal he may be, who is beyond redemption. There is no place at whose gates it is written: Abandon all hope, ye who enter here. There are no individuals who are utterly evil. Their characters have to be understood from within the context of their lives. Perhaps the criminals are diseased fellow-men whose love has lost its proper aim.

All men are the children of the Immortal, *amṛtasya putrāḥ.* The spirit is in everyone as a part of one's self, as a part of the very substratum of one's being. It may be buried in some like a hidden treasure beneath a barren debris of brutality and violence — but it is there all the same, operative and alive, ready to come to the surface at the first suitable opportunity. 'The light which lighteth every man that cometh into the world cannot be put out.' Whether we like it or not, whether we know it or not, the Divine is in us and the end of man consists in attaining conscious union with the Divine. A Japanese Zen Buddhist teacher observes: 'There is no hamlet so forlorn that the rays of the silver moon fail to reach it. Nor is there any man who by opening wide the windows of his thought cannot perceive divine truth and take it into his heart.'

The distinction between the kingdoms of light and of darkness, between heaven and hell becomes untenable. The cosmic power of the Eternal, His universal love will not suffer defeat. Hindu and Buddhist systems aim at universal salvation. According to *Mahāyāna* Buddhism, the Buddha deliberately refrained from coming to the final term of enlightenment in order to help others on the way. He has taken a vow that he will not enter into nirvāṇa until everything that exists, every particle of corruptible dust, has reached the goal.

This does not mean that the Hindu and the Buddhist religions cancel the distinction between good and evil. It only means that even the evil have other chances. The Divine provides the soul with a succession of spiritual opportunities. If there is only one chance given to human beings, they have at the end of this one life to be redeemed if good, or condemned if evil. Such a doctrine is not consistent with the view that God is infinite love, infinite compassion. India has stood for an ideal that does not make man merely a creature of time, dependent solely on his material conditions and possessions, and confined to them. We have proclaimed that the world is under moral law, that the life is the scene of man's moral choice. It is *dharma-kṣetra.* It is never too late for man to strive and attain his full stature. For the Hindu and the Buddhist, religion is a transforming experience. It is not a theory of God; it is spiritual consciousness, insight into Reality. Belief and conduct, rites and ceremonies, dogmas and authorities are subordinate to the art of conscious self-discovery and contact with the Divine. When the individual withdraws his soul from all outward events, gathers himself together inwardly, strives with concentration, there dawns upon him an experience sacred, strange, wondrous, which quickens within him, lays hold on him, becomes his very being. The possibility of this experience constitutes the most conclusive proof of the reality of God. Even those who are the children of science and reason must submit to the fact of spiritual experience which is primary and positive. We may dispute theologies, but we cannot deny facts. The fire of life in its visible burning compels assent, though not the fumbling speculations of smokers sitting around the fire.

While realization is a fact, the theory of reality is an inference. There is a difference between contact with reality and opinion about it, between the mystery of godliness and belief in God.

Rationalistic self-sufficiency is dangerous. The human mind is sadly crippled in its religious thinking by the belief that truth has been found, embodied, standardized, and nothing remains for man but to reproduce in his feebleness some treasured feature of an immutable perfection which is distant from him. Claims to infallible truth, based on alleged revelations, are not compatible with religion as spiritual adventure. The fulfilment of man's life is spiritual experience in which every aspect of man's being is raised to its highest point; all the senses gather, the whole mind leaps forward and realizes in one quivering instant such things as cannot be expressed. Though it is beyond the world of tongue or concept of mind, the longing and love of the soul, its desire and anxiety, its seeking and thinking are filled with the highest spirit. This is religion. It is not mere argument about it.

* * *

When we frame theories of religion, we turn the being of the soul into the having of a thing. We transform what originally comprehended our being into some object which we ourselves comprehend. Thus the total experience becomes an item of knowledge. Our disputes about dogmas are in regard to these partial items of knowledge. At its depth, religion in its silences and expressions is the same. There is a common ground on which the different religious traditions rest. This common ground belongs of right to all of us, as it has its source in the non-historical, the eternal; the universality of fundamental ideas which historical studies demonstrate is the hope of the future. It will make for religious unity and cultural understanding. The essential points of the Asian outlook on life, which are also to be found in the great tradition of spiritual life in the West, give us the basic certainties for the new world which is on the horizon. These are the divine possibilities of the soul: faith in democracy, unity of all life and existence, insistence on the active reconciliation of different faiths and cultures so as to promote the unity of mankind.

Modern civilization, which is becoming increasingly technological, tends to concentrate on a limited order of truth. It accepts the scientifically verifiable as the only basis for action. Some scientists and technicians who have emerged as the leaders of our age, speak of man as a purely mechanical material being, a creature made up of automatic reflexes. They emphasize the more earthly propensities of men and women. They seem to be blind to the higher sanctity which lives in man. Those who are born in this age feel the loss of faith; they are the spiritually displaced; they are the culturally uprooted; they are the traditionless. The only hope for man is a spiritual recovery, the realization that he is an unfinished animal and his goal is the Kingdom of God which is latent in him.

> All epochs dominated by belief, in whatever shape, have a radiance and bliss of their own and bear fruit for their people as well as for posterity. All epochs over which unbelief, in whatever form, maintains its miserable victory are ignored by posterity, because nobody likes to tug his life out over sterile things.

Few people would deny the truth of this statement of Goethe or that this is an age of unbelief. It is an age not so much unlit by belief, as lacking the very capacity to believe. The modern community, as a community, has lost its sense of the relatedness of things. There is a void today in men's minds which dogmatic religions are unable to fill. When the old gods, the old verities, the old values are fading, when life itself has become dim and its very forms are stiffening, there are always some intense natures to whom it is intolerable that there should not already be new and greater faiths in sight. We are too profoundly religious to be able to endure this precarious predicament.

When Graeco-Roman civilization was triumphant, it failed to supply its conquered people with a religion and, instead, was itself conquered by a religion supplied by them. May it not be that today the peoples of Asia may supply a spiritual orientation to the new world based on science and technology? By its material and political devices, the West is able to provide a secure framework of order within which different civilizations could mingle, and fruitful intercourse between them can take place by which the spiritual poverty of the world can be overcome. Without a spiritual recovery, the scientific achievement threaten to destroy us. We are living in days of destiny. Either the world will blow up in flames or settle down in peace. It depends on the seriousness with which we face the tasks of our age. A human society worthy of our science and the mobilized wisdom of the world can be built if those in power and position are willing to submit to severities which are not so drastic as a war will demand.

We have to reckon with the spirit of science, understand its limitations and develop an outlook which is consistent with its findings. Science will triumph over ignorance and superstition, and religion over selfishness and fear.

CHAPTER 7

SCIENCE AND RELIGION

We have travelled, across the centuries, from the stone age to the space age. Achievements of science and technology so far, are among the greatest works of reason. This technological revolution has spread over the whole world. The food we eat, the clothes we wear, the houses we live in, the words we use, the thoughts we think, and the way we entertain ourselves have all been produced by a large number of industrial processes, large-scale industry, mass production and labour-saving devices. These have made possible, a great improvement in standards of living and have contributed to the comfort of life. In our fight against disease, poverty, hunger, we can use the resources placed at our disposal by science. We can change living conditions for the better. In the London *Times*, there was a report from Stockholm of a brain surgery 'in which for the first time a beam of protons was used, instead of surgical instruments. Not even the skin of the skull had to be pierced and not a drop of blood was spilt...During the treatment the patient is kept "rotating" so that the beam which is produced by a big syncrocyclotrone and goes straight through the tissue to a depth of eight inches, will hit the affected area of the brain from several different angles. Only the exact spot on which the beam is focused — on this occasion it was a fingernail size — is "burnt away".' Such are the marvels of modern science which help us to enjoy this wonderful world into which we are born.

Many people who live in this science-dominated society claim that scientific knowledge would bring with it perpetual progress, a steady improvement in human relations. The expected transformation of men and of their social relationships has not been achieved. This period of great scientific achievement has also seen increased human misery, two World Wars, concentration camps, nuclear destruction and now terrorism which may grow into another war through accident or design. Growth in human wisdom has not been commensurate with the increase in scientific knowledge and technological power. The fear of universal destruction hangs over us like a dark cloud. Many people feel that the lights are failing and the shadows are growing darker with the increase of nuclear and biological armaments. There is a feeling of disintoxication, of disenchantment, of anxiety, even despair. All these are symptoms of man's continual conflict with himself. Science has liberated man from much of the tyranny of the environment but has not freed him from the tyranny of his own nature.

* * *

The sources of human happiness and social co-operation are not exactly the same as those of scientific inquiry. For the proper adjustment of man to the new world, an education of the human spirit is essential. To remake society, we have to remake ourselves. Humanities which cover art and literature, philosophy and religion, are as important for human welfare as science and technology. The two are not antagonistic to each other. Both in India and the West, science and religion had a common origin. The seer and the scientist were the same in the *Vedic āśrama* and in the Pythagorean brotherhood. Science itself was called 'natural philosophy' and its history is an essential part of the spiritual history of mankind. Science and technology on the one side, ethics and religion on the other, were sundered in later stages thus creating the problem of faith *vs.* reason, ethics *vs.* technics. The conflict between the two is a symptom of the split consciousness which is so characteristic of the mental disorder of the day. The question is often asked, whether we can preserve our ethical and spiritual values in an increasingly technological civilization.

Our age is the age of the specialist. Each one knows more and more about less and less. We concentrate on some narrow field and forget the larger context in which we can see the meaning of our own specialism. Modern specialization has led to the fragmentation of knowledge. We should not only be specialists but also have a sense of the meaning of life and of social responsibility. We have to reckon with the spirit of science, understand its limitations and develop an outlook which is consistent with its findings. It is no use clinging to traditional forms which have lost their meaning. We cannot ignore the world of scientific achievement and withdraw into the inner life of contemplation. We are involved in the mechanism of the modern world and so should seek even religious truth not merely with our emotions but with our minds. We cannot go back on the scientific civilization, nor can we drop religion. To reconcile the two is the task set to our generation.

Let us look at some of the difficulties in the present situation. Philosophy itself is dominated by the spirit of science. Positivism or logical empiricism aims at reducing all knowledge to foundations in experience construed as what is given or presented. It adopts the verificability theory of meaning by which the meaning of a sentence is to be determined by the specification of the manner in which its truth might be tested. Concepts are to be reducible to presented data. Those which cannot be reduced to given data are dismissed as meaningless. Traditional philosophy is replaced by purely epistemological question, syntactical and semantic. Philosophy is essentially analysis of language and the logical clarification of thoughts.

The authoritarian habit of mind is inconsistent with the empiricism of modern science. The scientist holds that only empirical, verifiable evidence is to be treated as true. Basic assertions about the nature of the universe, of life and death do not carry conviction to those trained in science.

Naturalists make out that man's soul is but 'a puff of vapour'. It is something thrown up by the random collisions of particles in aimless flight. It is an accident and death, may blot out the human species. Man's mind is a material conglomeration of swiftly moving atoms. The nature of the human self is interpreted in a way which robs it of its reality. Biologists argue that the individual person is the product of his heredity and environment. Psychologists tell us that the subconscious mind, which is the seat of instincts and emotions is the determining factor in man's life. Sociologists argue that the social environment in which a man is born moulds his mind and character. He is ruled, fed, clothed and educated as a social unit. Even his hobbies and amusements are standardized. He becomes a mere function of society. He is deprived of the chance to live his own life. Man has become a slave of the machines, which have conquered space and time for him. We speak in terms of large numbers and mass organizations; the human being is de-individualized.

On such a materialist view, the role of the individual in the shaping of history is negligible. There is a force or principle that governs human affairs, historic destiny, fate, providence, history, economic laws. There is an inevitability about the future of man.

* * *

Modern science is not committed to the mechanistic materialist view. The cosmos which science studies is not adequately interpreted by naturalist science. The *Bhagavadgītā* tells us that we know only the middle, not the beginning or the end. Scientific statements are valid only within the framework of strictly limited relations. While they study the working of atoms and electrons, they do not and cannot say anything about the creation of atoms and electrons. The scientist who is a part of the universe, cannot as a scientist make any statement about the meaning of the universe as a whole. Max Planck says:

> 'Science cannot solve the ultimate mystery of nature. And that is because in the last analysis, we ourselves are part of nature and therefore part of the mystery we are trying to solve. The most penetrating eye cannot see itself any more than a working instrument can work upon itself.'[1]

---

[1] *Where is Science Going?*

Besides, science is primarily construction of theories which enable us to understand some domain of observationally ascertained fact. A theory is a pattern of ideas used to interpret the given data of experience. The scientist does not confine his attention to the things that are observed, but goes down to a deeper level and frames scientific entities, which are not capable of direct observation but necessary to account for the facts of observation. Philosophical theories arise out of reflections on the data, the universe has given to us. The *Braham Sūtra* opens with with the words, *athāto brahma jijñāsā.* Now, therefore, an enquiry into Brahman. The next *sutra* tells us that the Supreme Spirit is the basis of the whole world process, its origin, maintenance and dissolution. The underlying structure of reality is accessible to reason because it is the product of reason. Nature is not self-subsistent but owes its being to a transcendent principle whose mind is reflected, however feebly, in it. This world is God's world: *īśāvāsyam idaṁ sarvam.*

This finding that there is an ultimate principle which underlies and informs the cosmic process is not only inferred but it is also intuited. Experience is not limited to the facts of space and time. We have experience of laws of nature, of logical universals as also of Ultimate Reality.

The soul's direct contact in full awareness of Transcendent Reality is a basic postulate of both Eastern and Western thoughts. Plato sees in the soul a being of divine origin. In it is the longing for reunion with the Divine. St. Paul's Damascus experience is an illustration of spiritual certainty. Augustine's words are well known: 'My mind in the flash of a trembling glance came to Absolute Being. That which is.'[2] Many seers have testified to the

[2] *Confessions,* VII. 17, 23.

vision of beauty and meaning in the heart of things. H.G. Wells in *First and Last Things* wrote:

> At times in the silence of the night and in rare, lonely moments I come upon a sort of communion of myself and something which is not myself...These moments happen and they are supreme fact in my religious life. They are the crown of my religious experience.

Anyone who has the experience is seized by it and does not indulge in fruitless metaphysical or epistemological speculation. Absolute certainty bring its own evidence, and has no need of proofs. Its numinosity and its rationality mark it as authentic experience.

Man is not saved by metaphysics. Spiritual life involves a change of consciousness, of being, rebirth of the spirit, *metanoia.* Those who have the experience decline to define the nature of the Ultimate Reality, as the human mind cannot comprehend it. So they are hospitable to the varying accounts of its nature. They realize that philosophy is a school of wisdom and a school of wonder.

The universe into which we are born and where we may be contriving our death has developed this marvel of an enquiring mind which reflects divine creativity. Men are made in the image of God, a God who is not angered by neglect or placated by prayers. The wheels of his chariot turn unimpeded by pity or anger. Heredity and environment, the subconscious mind, society and its institutions are all facts but there is, in addition, the will of man which brings something new and different. Freedom is the meaning of human reality. Through our choice, a new event is produced, unique and real. Included in human nature is the capacity to go beyond our heritage, biological, psychological and sociological. Human choice is not absolutely uncaused and unconditioned. The creative principle is functioning in the mind of man. The human individual is the source of liberty. It is something that escapes scientific or objective analysis. Every great scientific discovery is a creative act. It is not a mere intellectual process, but a moral achievement. It is the work of one who brings to his task great effort and self-sacrifice. If science tells us anything, it is this, that the human individual is an active agent, a doer, a creator. He is not a mere pawn of fate, a slave of necessity.

* * *

The greatest thinkers in the East and the West have felt the need for both science and spiritual life. Civilization is intended to make us aware of the creativity in us. The freedom of the individual should not be reduced to physical and spiritual slavery. The function of civilization is not to produce robots who carry out the directions of any mad dictator. As members of groups we do things which we will not allow ourselves to do in our private lives. Man has shown himself capable of indescribable horror. This is because man, the lord of all the elements, — earth, air and water, hugs to his bosom notions which diminish his dignity and reduce his freedom to absurdity. Man the inventor and vehicle of all the great developments of science and technology has become a problem, an enigma to himself. The problem can be solved only when we turn to the inner life of the individual. Immediate relation to Ultimate Reality is the basis of certainty which helps the individual from dissolving in the crowd. He will not then be swept by the drift of things. He will then choose that the creations of the human mind shall be a blessing and not a curse to mankind. If fear and anxiety are mixed with hope and assurance, it is because our natures have not become integrated.

The world is in tumult. It is wrong to think that no nation will embark on a war unless it has a reasonable chance of winning it. Since a nuclear war is obviously suicidal to all those who engage in it, nobody will start it. Such an assumption presupposes the role of reason in human affairs which is quite doubtful. What rules events in the world is not logic but human nature. It is full of passions and is not deterred by the fear of consequences. People lead harassed lives where compulsive emotions are so powerful that there is no scope for rational thought or calm deliberation. There is a conflict within us between necessity and freedom, between hate and love, death and life. The world is a battlefield and what is required of us is to be not on the winning side but on the right side.

Century after century of this restless planet's history has thrown up challenges. History is a perpetual process of change. It is always taking away old things and bringing new things into being. Some things last longer than others but none lasts for ever. The speed of the rocket and the power of the hydrogen bomb make out that we face a common challenge and so should share a common cause. History must now change its course. We must work for life, conquering the impulses of darkness and death in us. It is a time for decision. If we are not to betray the human race we must change and grow in greatness. The alternatives are not race suicide by accident or race slavery by design. There can be education of man and friendship among nations. We must give up ideological intolerance. In the present state of muddled thinking the best in the human tradition has to be preserved. In an essay, *Of Innovations* Francis Bacon writes:

> And he that will not apply new remedies must expect new evils; for time is the greatest innovator; and if time, of

course, alters things to the worse and wisdom and counsel shall not alter them to the better, what shall be the end?

We must subdue the passions of greed, hate and selfishness and attain self-conquest and self-dedication.

Under the influence of the universal experiences of religions and modern science and technology, mankind is being moulded into a single community. The common man has to discover what is uncommon in him and learn to live as a citizen of a world community. Science will triumph over ignorance and superstition, and religion over selfishness and fear, and nations will come together to build a great future for humanity, the brotherhood of man which has been the vision of the prophets since the beginning of time.

Also in Orient Paperbacks

The Bhagvadgita

M.K. Gandhi

In Mahatma Gandhi's own words his interpretation of the *Bhagvadgita* is designed for the common man—'who has little or no literary equipment, who has neither the time nor the desire to read the *Gita* in the original, and yet who stands in need of its support.'

'Mahatma Gandhi's interpretation is unique by virtue of the simple style and illustrations from practical life...makes an interesting reading...'

Hindustan Times

'Designed for the common man who has no time or intellectual equipment to read the Gita in original...'

M.P. Chronicle

'I regard Gandhi as the only truly great figure of our age.'

Albert Einstein

Also in Orient Paperbacks

Scholar Extraordinary:
The Life of Friedrich Max Muller
Nirad C Chaudhuri

'This book is a biography, an account of the life of a man who was a scholar and a thinker...who played so important and significant a role in history that he remains an element to be reckoned with in understanding the continuing evolution of a particular people or humanity in general; his personality and activities belong to a type whose presence and functioning is continuous and universal, relevant to all ages'

From the Introduction